50% OFF Online SAT Prep Course!

Dear Customer,

We consider it an honor and a privilege that you chose our SAT Study Guide. As a way of showing our appreciation and to help us better serve you, we are offering **50% off their online SAT Prep Course.** Many SAT courses are needlessly expensive and don't deliver enough value. With their course, you get access to the best SAT prep material, and you only pay half price.

Mometrix has structured their online course to perfectly complement your printed study guide. The SAT Prep Course contains **in-depth lessons** that cover all the most important topics, **120+ video reviews** that explain difficult concepts, over **800 practice questions** to ensure you feel prepared, and more than **400 digital flashcards**, so you can study while you're on the go.

Online SAT Prep Course

Topics Covered:

- Reading
  - o Literary Analysis
  - o Inferences in a Text or Texts
  - o Author's Use of Language
- Writing
  - o Foundations of Grammar
  - o Essay Revision and Sentence Logic
  - o Agreement and Sentence Structure
- Mathematics
  - o Algebra and Functions
  - o Geometry and Measurement
  - o Data Analysis, Statistics, and Probability

Course Features:

- SAT Study Guide
  - o Get content that complements our best-selling study guide.
- Full-Length Practice Tests
  - o With over 800 practice questions, you can test yourself again and again.
- Mobile Friendly
  - o If you need to study on the go, the course is easily accessible from your mobile device.
- SAT Flashcards
  - o Our course includes a flashcards mode with over 400 content cards for you to study.

To receive this discount, visit their website: mometrix.com/university/sat and add the course to your cart. At the checkout page, enter the discount code: **TPBSAT50**

If you have any questions or concerns, please contact them at universityhelp@mometrix.com.

Sincerely,

 partnership with

FREE Test Taking Tips DVD Offer

To help us better serve you, we have developed a Test Taking Tips DVD that we would like to give you for FREE. **This DVD covers world-class test taking tips that you can use to be even more successful when you are taking your test.**

All that we ask is that you email us your feedback about your study guide. Please let us know what you thought about it – whether that is good, bad or indifferent.

To get your **FREE Test Taking Tips DVD**, email freedvd@studyguideteam.com with "FREE DVD" in the subject line and the following information in the body of the email:

a. The title of your study guide.

b. Your product rating on a scale of 1-5, with 5 being the highest rating.

c. Your feedback about the study guide. What did you think of it?

d. Your full name and shipping address to send your free DVD.

If you have any questions or concerns, please don't hesitate to contact us at freedvd@studyguideteam.com.

Thanks again!

SAT Prep 2021 and 2022 Book

SAT Study Guide with Practice Tests
[6th Edition]

TPB Publishing

Written and edited by TPB Publishing.

TPB Publishing is not associated with or endorsed by any official testing organization. TPB Publishing is a publisher of unofficial educational products. All test and organization names are trademarks of their respective owners. Content in this book is included for utilitarian purposes only and does not constitute an endorsement by TPB Publishing of any particular point of view.

Interested in buying more than 10 copies of our product? Contact us about bulk discounts:
bulkorders@studyguideteam.com

ISBN 13: 9781628456868
ISBN 10: 1628456868

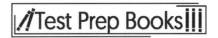

Table of Contents

Quick Overview

As you draw closer to taking your exam, effective preparation becomes more and more important. Thankfully, you have this study guide to help you get ready. Use this guide to help keep your studying on track and refer to it often.

This study guide contains several key sections that will help you be successful on your exam. The guide contains tips for what you should do the night before and the day of the test. Also included are test-taking tips. Knowing the right information is not always enough. Many well-prepared test takers struggle with exams. These tips will help equip you to accurately read, assess, and answer test questions.

A large part of the guide is devoted to showing you what content to expect on the exam and to helping you better understand that content. In this guide are practice test questions so that you can see how well you have grasped the content. Then, answer explanations are provided so that you can understand why you missed certain questions.

Don't try to cram the night before you take your exam. This is not a wise strategy for a few reasons. First, your retention of the information will be low. Your time would be better used by reviewing information you already know rather than trying to learn a lot of new information. Second, you will likely become stressed as you try to gain a large amount of knowledge in a short amount of time. Third, you will be depriving yourself of sleep. So be sure to go to bed at a reasonable time the night before. Being well-rested helps you focus and remain calm.

Be sure to eat a substantial breakfast the morning of the exam. If you are taking the exam in the afternoon, be sure to have a good lunch as well. Being hungry is distracting and can make it difficult to focus. You have hopefully spent lots of time preparing for the exam. Don't let an empty stomach get in the way of success!

When travelling to the testing center, leave earlier than needed. That way, you have a buffer in case you experience any delays. This will help you remain calm and will keep you from missing your appointment time at the testing center.

Be sure to pace yourself during the exam. Don't try to rush through the exam. There is no need to risk performing poorly on the exam just so you can leave the testing center early. Allow yourself to use all of the allotted time if needed.

Remain positive while taking the exam even if you feel like you are performing poorly. Thinking about the content you should have mastered will not help you perform better on the exam.

Once the exam is complete, take some time to relax. Even if you feel that you need to take the exam again, you will be well served by some down time before you begin studying again. It's often easier to convince yourself to study if you know that it will come with a reward!

Test-Taking Strategies

1. Predicting the Answer

When you feel confident in your preparation for a multiple-choice test, try predicting the answer before reading the answer choices. This is especially useful on questions that test objective factual knowledge. By predicting the answer before reading the available choices, you eliminate the possibility that you will be distracted or led astray by an incorrect answer choice. You will feel more confident in your selection if you read the question, predict the answer, and then find your prediction among the answer choices. After using this strategy, be sure to still read all of the answer choices carefully and completely. If you feel unprepared, you should not attempt to predict the answers. This would be a waste of time and an opportunity for your mind to wander in the wrong direction.

2. Reading the Whole Question

Too often, test takers scan a multiple-choice question, recognize a few familiar words, and immediately jump to the answer choices. Test authors are aware of this common impatience, and they will sometimes prey upon it. For instance, a test author might subtly turn the question into a negative, or he or she might redirect the focus of the question right at the end. The only way to avoid falling into these traps is to read the entirety of the question carefully before reading the answer choices.

3. Looking for Wrong Answers

Long and complicated multiple-choice questions can be intimidating. One way to simplify a difficult multiple-choice question is to eliminate all of the answer choices that are clearly wrong. In most sets of answers, there will be at least one selection that can be dismissed right away. If the test is administered on paper, the test taker could draw a line through it to indicate that it may be ignored; otherwise, the test taker will have to perform this operation mentally or on scratch paper. In either case, once the obviously incorrect answers have been eliminated, the remaining choices may be considered. Sometimes identifying the clearly wrong answers will give the test taker some information about the correct answer. For instance, if one of the remaining answer choices is a direct opposite of one of the eliminated answer choices, it may well be the correct answer. The opposite of obviously wrong is obviously right! Of course, this is not always the case. Some answers are obviously incorrect simply because they are irrelevant to the question being asked. Still, identifying and eliminating some incorrect answer choices is a good way to simplify a multiple-choice question.

4. Don't Overanalyze

Anxious test takers often overanalyze questions. When you are nervous, your brain will often run wild, causing you to make associations and discover clues that don't actually exist. If you feel that this may be a problem for you, do whatever you can to slow down during the test. Try taking a deep breath or counting to ten. As you read and consider the question, restrict yourself to the particular words used by the author. Avoid thought tangents about what the author *really* meant, or what he or she was *trying* to say. The only things that matter on a multiple-choice test are the words that are actually in the question. You must avoid reading too much into a multiple-choice question, or supposing that the writer meant something other than what he or she wrote.

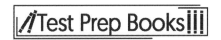

5. No Need for Panic

It is wise to learn as many strategies as possible before taking a multiple-choice test, but it is likely that you will come across a few questions for which you simply don't know the answer. In this situation, avoid panicking. Because most multiple-choice tests include dozens of questions, the relative value of a single wrong answer is small. As much as possible, you should compartmentalize each question on a multiple-choice test. In other words, you should not allow your feelings about one question to affect your success on the others. When you find a question that you either don't understand or don't know how to answer, just take a deep breath and do your best. Read the entire question slowly and carefully. Try rephrasing the question a couple of different ways. Then, read all of the answer choices carefully. After eliminating obviously wrong answers, make a selection and move on to the next question.

6. Confusing Answer Choices

When working on a difficult multiple-choice question, there may be a tendency to focus on the answer choices that are the easiest to understand. Many people, whether consciously or not, gravitate to the answer choices that require the least concentration, knowledge, and memory. This is a mistake. When you come across an answer choice that is confusing, you should give it extra attention. A question might be confusing because you do not know the subject matter to which it refers. If this is the case, don't eliminate the answer before you have affirmatively settled on another. When you come across an answer choice of this type, set it aside as you look at the remaining choices. If you can confidently assert that one of the other choices is correct, you can leave the confusing answer aside. Otherwise, you will need to take a moment to try to better understand the confusing answer choice. Rephrasing is one way to tease out the sense of a confusing answer choice.

7. Your First Instinct

Many people struggle with multiple-choice tests because they overthink the questions. If you have studied sufficiently for the test, you should be prepared to trust your first instinct once you have carefully and completely read the question and all of the answer choices. There is a great deal of research suggesting that the mind can come to the correct conclusion very quickly once it has obtained all of the relevant information. At times, it may seem to you as if your intuition is working faster even than your reasoning mind. This may in fact be true. The knowledge you obtain while studying may be retrieved from your subconscious before you have a chance to work out the associations that support it. Verify your instinct by working out the reasons that it should be trusted.

8. Key Words

Many test takers struggle with multiple-choice questions because they have poor reading comprehension skills. Quickly reading and understanding a multiple-choice question requires a mixture of skill and experience. To help with this, try jotting down a few key words and phrases on a piece of scrap paper. Doing this concentrates the process of reading and forces the mind to weigh the relative importance of the question's parts. In selecting words and phrases to write down, the test taker thinks about the question more deeply and carefully. This is especially true for multiple-choice questions that are preceded by a long prompt.

9. Subtle Negatives

One of the oldest tricks in the multiple-choice test writer's book is to subtly reverse the meaning of a question with a word like *not* or *except*. If you are not paying attention to each word in the question, you can easily be led astray by this trick. For instance, a common question format is, "Which of the following is…?" Obviously, if the question instead is, "Which of the following is not…?," then the answer will be quite different. Even worse, the test makers are aware of the potential for this mistake and will include one answer choice that would be correct if the question were not negated or reversed. A test taker who misses the reversal will find what he or she believes to be a correct answer and will be so confident that he or she will fail to reread the question and discover the original error. The only way to avoid this is to practice a wide variety of multiple-choice questions and to pay close attention to each and every word.

10. Reading Every Answer Choice

It may seem obvious, but you should always read every one of the answer choices! Too many test takers fall into the habit of scanning the question and assuming that they understand the question because they recognize a few key words. From there, they pick the first answer choice that answers the question they believe they have read. Test takers who read all of the answer choices might discover that one of the latter answer choices is actually *more* correct. Moreover, reading all of the answer choices can remind you of facts related to the question that can help you arrive at the correct answer. Sometimes, a misstatement or incorrect detail in one of the latter answer choices will trigger your memory of the subject and will enable you to find the right answer. Failing to read all of the answer choices is like not reading all of the items on a restaurant menu: you might miss out on the perfect choice.

11. Spot the Hedges

One of the keys to success on multiple-choice tests is paying close attention to every word. This is never truer than with words like almost, most, some, and sometimes. These words are called "hedges" because they indicate that a statement is not totally true or not true in every place and time. An absolute statement will contain no hedges, but in many subjects, the answers are not always straightforward or absolute. There are always exceptions to the rules in these subjects. For this reason, you should favor those multiple-choice questions that contain hedging language. The presence of qualifying words indicates that the author is taking special care with his or her words, which is certainly important when composing the right answer. After all, there are many ways to be wrong, but there is only one way to be right! For this reason, it is wise to avoid answers that are absolute when taking a multiple-choice test. An absolute answer is one that says things are either all one way or all another. They often include words like *every*, *always*, *best*, and *never*. If you are taking a multiple-choice test in a subject that doesn't lend itself to absolute answers, be on your guard if you see any of these words.

12. Long Answers

In many subject areas, the answers are not simple. As already mentioned, the right answer often requires hedges. Another common feature of the answers to a complex or subjective question are qualifying clauses, which are groups of words that subtly modify the meaning of the sentence. If the question or answer choice describes a rule to which there are exceptions or the subject matter is complicated, ambiguous, or confusing, the correct answer will require many words in order to be expressed clearly and accurately. In essence, you should not be deterred by answer choices that seem excessively long. Oftentimes, the author of the text will not be able to write the correct answer without

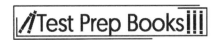

offering some qualifications and modifications. Your job is to read the answer choices thoroughly and completely and to select the one that most accurately and precisely answers the question.

13. Restating to Understand

Sometimes, a question on a multiple-choice test is difficult not because of what it asks but because of how it is written. If this is the case, restate the question or answer choice in different words. This process serves a couple of important purposes. First, it forces you to concentrate on the core of the question. In order to rephrase the question accurately, you have to understand it well. Rephrasing the question will concentrate your mind on the key words and ideas. Second, it will present the information to your mind in a fresh way. This process may trigger your memory and render some useful scrap of information picked up while studying.

14. True Statements

Sometimes an answer choice will be true in itself, but it does not answer the question. This is one of the main reasons why it is essential to read the question carefully and completely before proceeding to the answer choices. Too often, test takers skip ahead to the answer choices and look for true statements. Having found one of these, they are content to select it without reference to the question above. Obviously, this provides an easy way for test makers to play tricks. The savvy test taker will always read the entire question before turning to the answer choices. Then, having settled on a correct answer choice, he or she will refer to the original question and ensure that the selected answer is relevant. The mistake of choosing a correct-but-irrelevant answer choice is especially common on questions related to specific pieces of objective knowledge. A prepared test taker will have a wealth of factual knowledge at his or her disposal, and should not be careless in its application.

15. No Patterns

One of the more dangerous ideas that circulates about multiple-choice tests is that the correct answers tend to fall into patterns. These erroneous ideas range from a belief that B and C are the most common right answers, to the idea that an unprepared test-taker should answer "A-B-A-C-A-D-A-B-A." It cannot be emphasized enough that pattern-seeking of this type is exactly the WRONG way to approach a multiple-choice test. To begin with, it is highly unlikely that the test maker will plot the correct answers according to some predetermined pattern. The questions are scrambled and delivered in a random order. Furthermore, even if the test maker was following a pattern in the assignation of correct answers, there is no reason why the test taker would know which pattern he or she was using. Any attempt to discern a pattern in the answer choices is a waste of time and a distraction from the real work of taking the test. A test taker would be much better served by extra preparation before the test than by reliance on a pattern in the answers.

FREE DVD OFFER

Don't forget that doing well on your exam includes both understanding the test content and understanding how to use what you know to do well on the test. We offer a completely FREE Test Taking Tips DVD that covers world class test taking tips that you can use to be even more successful when you are taking your test.

All that we ask is that you email us your feedback about your study guide. To get your **FREE Test Taking Tips DVD**, email freedvd@studyguideteam.com with "FREE DVD" in the subject line and the following information in the body of the email:

- The title of your study guide.
- Your product rating on a scale of 1-5, with 5 being the highest rating.
- Your feedback about the study guide. What did you think of it?
- Your full name and shipping address to send your free DVD.

Introduction to the SAT

Function of the Test

The SAT is a standardized test taken by high school students across the United States and given internationally for college placement. It is designed to measure problem solving ability, communication, and understanding complex relationships. The SAT also serves as a qualifying measure to identify students for college scholarships, depending on the college being applied to. All colleges in the U.S. accept the SAT, and, in addition to admissions and scholarships, use SAT scores for course placement as well as academic counseling.

Most of the high school students who take the SAT are seniors. In 2016, the number of students who took the SAT was just under 1.7 million. It's important to note that since many updates have been implemented during the 2016 year, the data points cannot be compared to those in previous years. In 2014, 42.6 percent of students met the College Board's "college and career readiness" benchmark, and in 2015, 41.9 percent met this benchmark.

Test Administration

The SAT is offered on seven days throughout the year at schools throughout the United States. Internationally, the SAT is offered on five days throughout the year. There are thousands of testing centers worldwide. Test-takers can view the test centers in their area when they register for the test, or they can view testing locations at the College Board website, a not-for-profit that owns and publishes the SAT.

The SAT registration fee is $46, and the SAT with Essay registration fee is $60, although both of these have fee waivers available. Also note that students outside the U.S. may have to pay an extra processing fee. Additional fees include registering by phone, changing fee, late registration fee, or a waitlist fee. Test-takers may register four score reports for free up to nine days after the test. Any additional score reports cost $12, although fee waivers are available for this as well.

Test Format

The SAT gauges a student's proficiency in three areas: Reading, Mathematics, and Writing and Language. The reading portion of the SAT measures comprehension, requiring candidates to read multi-paragraph fiction and non-fiction segments including informational visuals, such as charts, tables and graphs, and answer questions based on this content. Fluency in problem solving, conceptual understanding of equations, and real-world applications are characteristics of the math test. The writing and language portion requires students to evaluate and edit writing and graphics to obtain an answer that correctly conveys the information given in the passage.

The SAT contains 154 multiple-choice questions, with each section comprising over 40 questions. A different length of time is given for each section, for a total of three hours, plus fifty minutes for the essay (optional).

Section	Time (In Minutes)	Number of Questions
Reading	65	52
Writing and Language	35	44
Mathematics	80	58
Essay (optional)	50 (optional)	1 (optional)
Total	**180**	**154 + optional essay**

Scoring

Scores for the new SAT are based on a scale from 400 to 1600. Scores range from 200 to 800 for Evidence-Based Reading and Writing, and 200 to 800 for Math. The optional essay is scored from 2 to 8. The SAT also no longer penalizes for incorrect answers. Therefore, a student's raw score is the number of correctly answered questions.

On the College Board website, there are indicators to determine what the benchmark scores are. The scores are divided up into green, yellow, or red. Green meets or exceeds the benchmark, and shows a 480 to 800 in Evidence-Based Reading and Writing, and a 530 to 800 in Math.

Recent/Future Developments

The SAT taken before March 2016 is different than the one administered currently. Currently, the essay is optional, and the time limit for the Reading, Writing, and Math sections has increased per section. The content features also vary, with the new test focusing on skills that research has identified as most important for college readiness and the meaning of words in extended context rather than emphasis on vocabulary. The score range has changed from 600–2400 to 400–1600. There has also been added subscore reporting, which provides insight to students and parents about the scores.

Study Prep Plan for the SAT Exam

1 **Schedule** - Use one of our study schedules below or come up with one of your own.

2 **Relax** - Test anxiety can hurt even the best students. There are many ways to reduce stress. Find the one that works best for you.

3 **Execute** - Once you have a good plan in place, be sure to stick to it.

Sample Study Plans

One Week Study Schedule

Day 1	Reading
Day 2	Writing and Language
Day 3	Mathematics
Day 4	Essay
Day 5	Practice Questions
Day 6	Review Answer Explanations
Day 7	Take Your Exam!

Two Week Study Schedule

Day 1	Command of Evidence	Day 8	Problem Solving and Data Analysis
Day 2	Words in Context	Day 9	Practice Questions
Day 3	Practice Questions	Day 10	Essay Prompt
Day 4	Expression of Ideas	Day 11	Review Practice Questions
Day 5	Standard English Conventions	Day 12	Review Answer Explanations
Day 6	Practice Questions	Day 13	(Study Break)
Day 7	Heart of Algebra	Day 14	Take Your Exam!

One Month Study Schedule

Day 1	Finding Evidence in a Passage	Day 11	Organization and Focus	Day 21	Four Types of Sentence Structures
Day 2	Author's Use of Evidence	Day 12	Introductions and Conclusions	Day 22	Heart of Algebra
Day 3	Informational Graphics	Day 13	Precision and Conciseness	Day 23	Probleming Solving and Data Analysis
Day 4	Using Context Clues	Day 14	Proposition and Support	Day 24	Determining Forms of Expressions
Day 5	Author's Word Choice Shapes Meaning	Day 15	Effective Language Use	Day 25	Additional Topics in Math
Day 6	Comprehending Test Questions	Day 16	Style, Tone, and Mood	Day 26	Analyzing the Essay
Day 7	Reading for Tone, Message, and Effect	Day 17	Syntax and Quantitative Information	Day 27	Essay Prompt
Day 8	Science Passages	Day 18	Types of Sentences	Day 28	Practice Questions
Day 9	Examine Hypotheses	Day 19	Parts of Speech	Day 29	Review Answer Explanations
Day 10	Interpreting Data and Considering Implications	Day 20	Independent and Dependent Clauses	Day 30	Take Your Exam!

Reading

The Reading section of the SAT measures test takers' ability to read, understand, and interpret the types of texts they will encounter in higher education and their vocational pursuits. The passages included in this section are drawn from previously-published literary or informational texts. Some passages include graphics such as charts, graphs, and tables, which the test taker must interpret alongside the written text to correctly answer the questions. Other questions will follow two passages that are purposely paired for the test taker to compare and contrast.

The Reading section contains 52 questions, which must be answered in the allotted 65 minutes. The majority of the passages in this section are pulled from informational or nonfiction works, covering topics pertaining to history, humanities, science, and social studies. At least one single passage or a pair of passages to be considered together will be drawn from the seminal founding U.S. documents or the *Great Global Conversation,* which is an umbrella term used to describe works from countries around the world and spanning several centuries that address universal topics like justice, freedom, and human dignity. Test takers will encounter only one literary passage in the Reading Section. This except is drawn from an American or international fictional piece, such as a published novel or short story.

In addition to encountering a variety of topics in the passages, test takers will need to employ their comprehension skills pertaining to understanding texts written for different purposes. Some texts may tell a story or describe a process. Others will present an argument and support for an opinion, while still other passages may delineate the purpose of a study or experiment. The text complexity or reading level of each passage also varies; passages may span the gamut from a ninth-grade reading level to that expected of a first-year university student or post-secondary scholar.

Questions on the Reading section will generally fall into one of three principle categories: information and ideas, rhetoric, and synthesis. As the name implies, **information and ideas** questions focus on the direct or indirect ideas or facts that the passage presents. The answers to some of these questions will not necessarily be explicitly revealed in the passage; rather, test takers must use their inferential skills to uncover the author's implied meaning. **Rhetoric** questions require test takers to consider the method or manner in which the author conveys his or her intended message. Lastly, **synthesis** questions ask test takers to draw conclusions from a passage or make connections between a pair of passages or between passages and an accompanying graphic.

While the vast majority of questions in the Reading section address the test taker's ability to understand written language and ideas in each passage, a small number of questions will also assess the test taker's ability to interpret graph, tables, and charts. For this reason, in addition to studying reading comprehension techniques and honing skills related to synthesis and inference, test takers should practice reading and interpreting tables, graphs, and charts. Test takers should review the material in this section pertaining to critical reading skills and refer to the information in the Writing and Language section about specific comprehension skills pertinent to history and the sciences texts, as well as graphic-interpreting techniques.

Command of Evidence

Command of evidence, or the ability to use contextual clues, factual statements, and corroborative phrases to support an author's message or intent, is an important part of the Reading section of the SAT. A test taker's ability to parse out factual information and draw conclusions based on evidence is important to critical reading comprehension. The test will ask students to read text passages, and then answer questions based on information contained in them. These types of questions may ask test takers to identify stated facts. They may also require test takers to draw logical conclusions, identify data based on graphs, make inferences, and to generally display analytical thinking skills.

Finding Evidence in a Passage

The basic tenet of reading comprehension is the ability to read and understand a text. One way to understand a text is to look for information that supports the author's main idea, topic, or position statement. This information may be factual, or it may be based on the author's opinion. This section will focus on the test taker's ability to identify factual information, as opposed to opinionated bias. The Reading section will ask test takers to read passages containing factual information, and then logically relate those passages by drawing conclusions based on evidence.

In order to identify factual information within a text passage, test takers should begin by looking for statements of fact. **Factual statements** can be either true or false. Identifying factual statements as opposed to opinion statements is important in demonstrating full command of evidence in reading. For example, the statement *The temperature outside was unbearably hot* may seem like a fact; however, it's not. While anyone can point to a temperature gauge as factual evidence, the statement itself reflects only an opinion. Some people may find the temperature unbearably hot. Others may find it comfortably warm. Thus, the sentence, *The temperature outside was unbearably hot,* reflects the opinion of the author who found it unbearable. If the text passage followed up the sentence with atmospheric conditions indicating heat indices above 140 degrees Fahrenheit, then the reader knows there is factual information that supports the author's assertion of *unbearably hot*.

In looking for information that can be proven or disproven, it's helpful to scan for dates, numbers, timelines, equations, statistics, and other similar data within any given text passage. These types of indicators will point to proven particulars. For example, the statement, *The temperature outside was unbearably hot on that summer day, July 10, 1913,* most likely indicates factual information, even if the reader is unaware that this is the hottest day on record in the United States. Be careful when reading biased words from an author. Biased words indicate opinion, as opposed to fact. The following list contains a sampling of common biased words:

- Good/bad
- Great/greatest
- Better/best/worst
- Amazing
- Terrible/bad/awful
- Beautiful/handsome/ugly
- More/most
- Exciting/dull/boring
- Favorite
- Very
- Probably/should/seem/possibly

Remember, most of what is written is actually opinion or carefully worded information that seems like fact when it isn't. To say, *duplicating DNA results is not cost-effective* sounds like it could be a scientific fact, but it isn't. Factual information can be verified through independent sources.

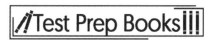

The simplest type of test question may provide a text passage, then ask the test taker to distinguish the correct factual supporting statement that best answers the corresponding question on the test. However, be aware that most questions may ask the test taker to read more than one text passage and identify which answer best supports an author's topic. While the ability to identify factual information is critical, these types of questions require the test taker to identify chunks of details, and then relate them to one another.

Displaying Analytical Thinking Skills

Analytical thinking involves being able to break down visual information into manageable portions in order to solve complex problems or process difficult concepts. This skill encompasses all aspects of command of evidence in reading comprehension.

A reader can approach analytical thinking in a series of steps. First, when approaching visual material, a reader should identify an author's thought process. Is the line of reasoning clear from the presented passage, or does it require inference and coming to a conclusion independent of the author? Next, a reader should evaluate the author's line of reasoning to determine if the logic is sound. Look for evidentiary clues and cited sources. Do these hold up under the author's argument? Third, look for bias. Bias includes generalized, emotional statements that will not hold up under scrutiny, as they are not based on fact. From there, a reader should ask if the presented evidence is trustworthy. Are the facts cited from reliable sources? Are they current? Is there any new factual information that has come to light since the passage was written that renders the argument useless? Next, a reader should carefully think about information that opposes the author's view. Do the author's arguments guide the reader to identical thoughts, or is there room for sound arguments? Finally, a reader should always be able to identify an author's conclusion and be able to weigh its effectiveness.

The ability to display analytical thinking skills while reading is key in any standardized testing situation. Test takers should be able to critically evaluate the information provided, and then answer questions related to content by using the steps above.

Making Inferences

Simply put, an **inference** is an educated guess drawn from evidence, logic, and reasoning. The key to making inferences is identifying clues within a passage, and then using common sense to arrive at a reasonable conclusion. Consider it "reading between the lines."

One way to make an inference is to look for main topics. When doing so, pay particular attention to any titles, headlines, or opening statements made by the author. Topic sentences or repetitive ideas can be clues in gleaning inferred ideas. For example, if a passage contains the phrase *While some consider DNA testing to be infallible, it is an inherently flawed technique,* the test taker can infer the rest of the passage will contain information that points to problems with DNA testing.

The test taker may be asked to make an inference based on prior knowledge but may also be asked to make predictions based on new ideas. For example, the test taker may have no prior knowledge of DNA other than its genetic property to replicate. However, if the reader is given passages on the flaws of DNA testing with enough factual evidence, the test taker may arrive at the inferred conclusion that the author does not support the infallibility of DNA testing in all identification cases.

When making inferences, it is important to remember that the critical thinking process involved must be fluid and open to change. While a reader may infer an idea from a main topic, general statement, or other clues, they must be open to receiving new information within a particular passage. New ideas presented by an author may require the test taker to alter an inference. Similarly, when asked questions that require making an inference, it's important to read the entire test passage and all of the answer options. Often, a test taker will need to refine a general inference based on new ideas that may be presented within the test itself.

Author's Use of Evidence to Support Claims

Authors utilize a wide range of techniques to tell a story or communicate information. Readers should be familiar with the most common of these techniques. Techniques of writing are also commonly known as rhetorical devices, and they are some of the evidence that authors use to support claims.

In nonfiction writing, authors employ argumentative techniques to present their opinion to readers in the most convincing way. Persuasive writing usually includes at least one type of appeal: an appeal to logic (**logos**), emotion (**pathos**), or credibility and trustworthiness (**ethos**). When a writer appeals to logic, they are asking readers to agree with them based on research, evidence, and an established line of reasoning. An author's argument might also appeal to readers' emotions, perhaps by including personal stories and **anecdotes** (a short narrative of a specific event). A final type of appeal, appeal to authority, asks the reader to agree with the author's argument on the basis of their expertise or credentials. Consider three different approaches to arguing the same opinion:

Logic (Logos)
Below is an example of an appeal to logic. The author uses evidence to disprove the logic of the school's rule (the rule was supposed to reduce discipline problems, but the number of problems has not been reduced; therefore, the rule is not working) and call for its repeal.

> Our school should abolish its current ban on cell phone use on campus. This rule was adopted last year as an attempt to reduce class disruptions and help students focus more on their lessons. However, since the rule was enacted, there has been no change in the number of disciplinary problems in class. Therefore, the rule is ineffective and should be done away with.

Emotion (Pathos)
An author's argument might also appeal to readers' emotions, perhaps by including personal stories and anecdotes.

The next example presents an appeal to emotion. By sharing the personal anecdote of one student and speaking about emotional topics like family relationships, the author invokes the reader's empathy in asking them to reconsider the school rule.

> Our school should abolish its current ban on cell phone use on campus. If they aren't able to use their phones during the school day, many students feel isolated from their loved ones. For example, last semester, one student's grandmother had a heart attack in the morning. However, because he couldn't use his cell phone, the student didn't know about his grandmother's accident until the end of the day—when she had already passed away, and it was too late to say goodbye. By preventing students from contacting their friends and family, our school is placing undue stress and anxiety on students.

All three examples begin from the same opinion—the school's phone ban needs to change—but rely on different argumentative styles to persuade the reader.

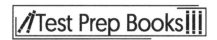

Credibility (Ethos)

Finally, an appeal to authority includes a statement from a relevant expert. In this case, the author uses a doctor in the field of education to support the argument.

> Our school should abolish its current ban on cell phone use on campus. According to Dr. Bartholomew Everett, a leading educational expert, "Research studies show that cell phone usage has no real impact on student attentiveness. Rather, phones provide a valuable technological resource for learning. Schools need to learn how to integrate this new technology into their curriculum." Rather than banning phones altogether, our school should follow the advice of experts and allow students to use phones as part of their learning.

Informational Graphics

A test taker's ability to draw conclusions from an informational graphic is a sub-skill in displaying one's command of reading evidence. Drawing **conclusions** requires the reader to consider all information provided in the passage, then to use logic to piece it together to form a reasonably correct resolution. In this case, a test taker must look for facts as well as opinionated statements. Both should be considered in order to arrive at a conclusion. These types of questions test one's ability to conduct logical and analytical thinking.

Identifying data-driven evidence in informational graphics is very similar to analyzing factual information. However, it often involves the use of graphics in order to do so. In these types of questions, the test taker will be presented with a graph, or organizational tool, and asked questions regarding the information it contains. On the following page, review the pie chart organizing percentages of primary occupations of public transportation passengers in US cities.

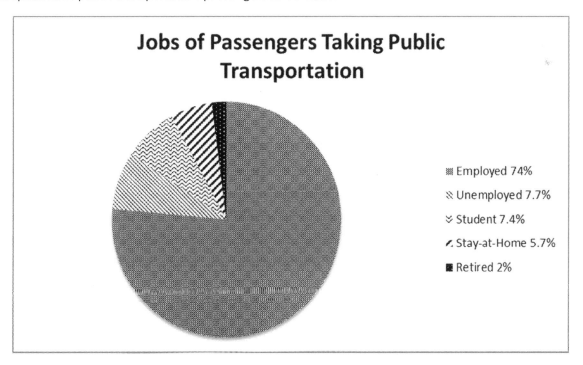

This figure depicts the jobs of passengers taking public transportation in U.S. cities. A corresponding SAT question may have the test taker study the chart, then answer a question regarding the values. For example, is the number of students relying on public transportation greater or less than the number of the unemployed? Similarly, the test may ask if people employed outside the home are less likely to use public transportation than homemakers. Note that the phrase *less likely* may weigh into the reader's choice of optional answers and that the test taker should look for additional passage data to arrive at a conclusion one way or another.

Words in Context

In order to successfully complete the Reading section of the SAT, the test taker should be able to identify the meaning of words in context. This involves a set of skills that requires the test taker to answer questions about unfamiliar words within a particular text passage. Additionally, the test taker may be asked to answer critical thinking questions based on unfamiliar word meanings. Identifying the meaning of different words in context is very much like solving a puzzle. By using a variety of techniques, a test taker should be able to correctly identify the meaning of unfamiliar words and concepts with ease.

Using Context Clues

A context clue is a hint that an author provides to the reader in order to help define difficult or unique words. When reading a passage, a test taker should take note of any unfamiliar words, and then examine the sentence around them to look for clues to the word meanings. Let's look at an example:

> He faced a *conundrum* in making this decision. He felt as if he had come to a crossroads. This was truly a puzzle, and what he did next would determine the course of his future.

The word *conundrum* may be unfamiliar to the reader. By looking at context clues, the reader should be able to determine its meaning. In this passage, context clues include the idea of making a decision and of being unsure. Furthermore, the author restates the definition of conundrum in using the word *puzzle* as a synonym. Therefore, the reader should be able to determine that the definition of the word *conundrum* is a difficult puzzle.

Similarly, a reader can determine difficult vocabulary by identifying antonyms. Let's look at an example:

> Her *gregarious* nature was completely opposite of her twin's, who was shy, retiring, and socially nervous.

The word *gregarious* may be unfamiliar. However, by looking at the surrounding context clues, the reader can determine that *gregarious* does not mean shy. The twins' personalities are being contrasted. Therefore, *gregarious* must mean sociable, or something similar to it.

At times, an author will provide contextual clues through a cause and effect relationship. Look at the next sentence as an example:

> The athletes were *elated* when they won the tournament; unfortunately, their off-court antics caused them to forfeit the win.

The word *elated* may be unfamiliar to the reader. However, the author defines the word by presenting a cause and effect relationship. The athletes were so elated at the win that their behavior went overboard, and they had to forfeit. In this instance, *elated* must mean something akin to overjoyed, happy, and overexcited.

Cause and effect is one technique authors use to demonstrate relationships. A **cause** is why something happens. The **effect** is what happens as a result. For example, a reader may encounter text such as *Because he was unable to sleep, he was often restless and irritable during the day.* The cause is insomnia due to lack of sleep. The effect is being restless and irritable. When reading for a cause and effect relationship, look for words such as "if", "then", "such", and "because." By using cause and effect, an author can describe direct relationships, and convey an overall theme, particularly when taking a stance on their topic.

An author can also provide contextual clues through **comparison and contrast**. Let's look at an example:

> Her torpid state caused her parents, and her physician, to worry about her seemingly sluggish well-being.

The word *torpid* is probably unfamiliar to the reader. However, the author has compared *torpid* to a state of being and, moreover, one that's worrisome. Therefore, the reader should be able to determine that *torpid* is not a positive, healthy state of being. In fact, through the use of comparison, it means sluggish. Similarly, an author may contrast an unfamiliar word with an idea. In the sentence *Her torpid state was completely opposite of her usual, bubbly self,* the meaning of *torpid*, or sluggish, is contrasted with the words *bubbly self*.

A test taker should be able to critically assess and determine unfamiliar word meanings through the use of an author's context clues in order to fully comprehend difficult text passages.

Relating Unfamiliar Words to Familiar Words

The Reading section of the SAT will test a reader's ability to use context clues, and then relate unfamiliar words to more familiar ones. Using the word *torpid* as an example, the test may ask the test taker to relate the meaning of the word to a list of vocabulary options and choose the more familiar word as closest in meaning. In this case, the test may say something like the following:

> Which of the following words means the same as the word *torpid* in the above passage?

Then they will provide the test taker with a list of familiar options such as happy, disgruntled, sluggish, and animated. By using context clues, the reader has already determined the meaning of *torpid* as slow or sluggish, so the reader should be able to correctly identify the word *sluggish* as the correct answer.

One effective way to relate unfamiliar word meanings to more familiar ones is to substitute the provided word in each answer option for the unfamiliar word in question. Although this will not always lead to a correct answer every time, this strategy will help the test taker narrow answer options. Be careful when utilizing this strategy. Pay close attention to the meaning of sentences and answer choices because it's easy to mistake answer choices as correct when they are easily substituted, especially when they are the same part of speech. Does the sentence mean the same thing with the substituted word option in place or does it change entirely? Does the substituted word make sense? Does it possibly mean the same as the unfamiliar word in question?

How an Author's Word Choice Shapes Meaning, Style, and Tone

Authors choose their words carefully in order to artfully depict meaning, style, and tone, which is most commonly inferred through the use of adjectives and verbs. The **tone** is the predominant emotion present in the text, and represents the attitude or feelings that an author has towards a character or event.

To review, an **adjective** is a word used to describe something, and usually precedes the noun, a person, place, or object. A **verb** is a word describing an action. For example, the sentence "The scary woodpecker ate the spider" includes the adjective "scary," the noun "woodpecker," and the verb "ate." Reading this sentence may rouse some negative feelings, as the word "scary" carries a negative charge. The **charge** is the emotional connotation that can be derived from the adjectives and verbs and is either positive or negative. Recognizing the charge of a particular sentence or passage is an effective way to understand the meaning and tone the author is trying to convey.

Many authors have conflicting charges within the same text, but a definitive tone can be inferred by understanding the meaning of the charges relative to each other. It's important to recognize key conjunctions, or words that link sentences or clauses together. There are several types and subtypes of conjunctions. Three are most important for reading comprehension:

1. **Cumulative conjunctions** add one statement to another.
   - Examples: and, both, also, as well as, not only
   - e.g. The juice is sweet *and* sour.
2. **Adversative conjunctions** are used to contrast two clauses.
   - Examples: but, while, still, yet, nevertheless
   - e.g. She was tired, *but* she was happy.
3. **Alternative conjunctions** express two alternatives.
   - Examples: or, either, neither, nor, else, otherwise
   - e.g. He must eat, *or* he will die.

Identifying the meaning and tone of a text can be accomplished with the following steps:

- Identify the adjectives and verbs.
- Recognize any important conjunctions.
- Label the adjectives and verbs as positive or negative.
- Understand what the charge means about the text.

To demonstrate these steps, examine the following passage from the classic children's poem, "The Sheep":

Lazy sheep, pray tell me why

In the pleasant fields you lie,

Eating grass, and daisies white,

From the morning till the night?

Everything can something do,

But what kind of use are you?

–Taylor, Jane and Ann. "The Sheep."

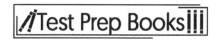

This selection is a good example of conflicting charges that work together to express an overall tone. Following the first two steps, identify the adjectives, verbs, and conjunctions within the passage. For this example, the adjectives are <u>underlined</u>, the verbs are in **bold**, and the conjunctions *italicized*:

> <u>Lazy</u> sheep, pray **tell** me why
>
> In the <u>pleasant</u> fields you **lie**,
>
> **Eating** grass, and daisies <u>white,</u>
>
> From the morning till the night?
>
> Everything can something do,
>
> *But* what kind of use are you?

For step three, read the passage and judge whether feelings of positivity or negativity arose. Then assign a charge to each of the words that were outlined. This can be done in a table format, or simply by writing a + or − next to the word.

The word <u>lazy</u> carries a negative connotation; it usually denotes somebody unwilling to work. To **tell** someone something has an exclusively neutral connotation, as it depends on what's being told, which has not yet been revealed at this point, so a charge can be assigned later. The word <u>pleasant</u> is an inherently positive word. To **lie** could be positive or negative depending on the context, but as the subject (the sheep) is lying in a pleasant field, then this is a positive experience. **Eating** is also generally positive.

After labeling the charges for each word, it might be inferred that the tone of this poem is happy and maybe even admiring or innocuously envious. However, notice the adversative conjunction, "but" and what follows. The author has listed all the pleasant things this sheep gets to do all day, but the tone changes when the author asks, "What kind of use are you?" Asking someone to prove their value is a rather hurtful thing to do, as it implies that the person asking the question doesn't believe the subject has any value, so this could be listed under negative charges. Referring back to the verb **tell**, after reading the whole passage, it can be deduced that the author is asking the sheep to tell what use the sheep is, so this has a negative charge.

+	−
Pleasant • Lie in fields • From morning to night	• Lazy • Tell me • What kind of use are you

Upon examining the charges, it might seem like there's an even amount of positive and negative emotion in this selection, and that's where the conjunction "but" becomes crucial to identifying the tone. The conjunction "but" indicates there's a contrasting view to the pleasantness of the sheep's daily life, and this view is that the sheep is lazy and useless, which is also indicated by the first line, "lazy sheep, pray tell me why."

It might be helpful to look at questions pertaining to tone. For this selection, consider the following question:

The author of the poem regards the sheep with a feeling of what?
    a. Respect
    b. Disgust
    c. Apprehension
    d. Intrigue

Considering the author views the sheep as lazy with nothing to offer, Choice *A* appears to reflect the opposite of what the author is feeling.

Choice *B* seems to mirror the author's feelings towards the sheep, as laziness is considered a disreputable trait, and people (or personified animals, in this case) with unfavorable traits might be viewed with disgust.

Choice *C* doesn't make sense within context, as laziness isn't usually feared.

Choice *D* is tricky, as it may be tempting to argue that the author is intrigued with the sheep because they ask, "pray tell me why." This is another out-of-scope answer choice as it doesn't *quite* describe the feelings the author experiences and there's also a much better fit in Choice *B*.

Analysis in History/Social Studies and in Science

The Reading section of the SAT will include one passage from social sciences such as history, economics, psychology, or sociology. For these types of passages and questions, the test taker will need to utilize all their reading comprehension skills, but mastery of further skills will help. This section addresses those skills.

Comprehending Test Questions Prior to Reading

While preparing for a historical passage on a standardized test, first read the test questions, and then quickly scan the test answers prior to reading the passage itself. Notice there is a difference between the terms **read** and **scan**. **Reading** involves full concentration while addressing every word. **Scanning** involves quickly glancing at text in chunks, noting important dates, words, and ideas along the way. Reading test questions will help the test taker know what information to focus on in the historical passage. Scanning answers will help the test taker focus on possible answer options while reading the passage.

When reading standardized test questions that address historical passages, be sure to clearly understand what each question is asking. Is a question asking about vocabulary? Is another asking for the test taker to find a specific historical fact? Do any of the questions require the test taker to draw conclusions, identify an author's topic, tone, or position? Knowing what content to address will help the test taker focus on the information they will be asked about later. However, the test taker should approach this reading comprehension technique with some caution. It is tempting to only look for the right answers within any given passage. However, do not put on "reading blinders" and ignore all other information presented in a passage. It is important to fully read every passage and not just scan it. Strictly looking for what may be the right answers to test questions can cause the test taker to ignore important contextual clues that actually require critical thinking in order to identify correct answers. Scanning a passage for what appears to be wrong answers can have a similar result.

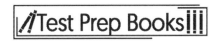

When reading test questions prior to tackling a historical passage, be sure to understand what skills the test is assessing, and then fully read the related passage with those skills in mind. Focus on every word in both the test questions and the passage itself. Read with a critical eye and a logical mind.

Reading for Factual Information

Standardized test questions that ask for factual information are usually straightforward. These types of questions will either ask the test taker to confirm a fact by choosing a correct answer, or to select a correct answer based on a negative fact question.

For example, the test taker may encounter a passage from Lincoln's Gettysburg address. A corresponding test question may ask the following:

> Which war is Abraham Lincoln referring to in the following passage?: "Now we are engaged in a great civil war, testing whether that nation, or any nation so conceived and so dedicated, can long endure."

This type of question is asking the test taker to confirm a simple fact. Given options such as World War I, the War of Spanish Succession, World War II, and the American Civil War, the test taker should be able to correctly identify the American Civil War based on the words "civil war" within the passage itself, and, hopefully, through general knowledge. In this case, reading the test question and scanning answer options ahead of reading the Gettysburg address would help quickly identify the correct answer. Similarly, a test taker may be asked to confirm a historical fact based on a negative fact question. For example, a passage's corresponding test question may ask the following:

> Which option is incorrect based on the above passage?

Given a variety of choices speaking about which war Abraham Lincoln was addressing, the test taker would need to eliminate all correct answers pertaining to the American Civil War and choose the answer choice referencing a different war. In other words, the correct answer is the one that contradicts the information in the passage.

It is important to remember that reading for factual information is straightforward. The test taker must distinguish fact from bias. **Factual statements** can be proven or disproven independent of the author and from a variety of other sources. Remember, successfully answering questions regarding factual information may require the test taker to re-read the passage, as these types of questions test for attention to detail.

Reading for Tone, Message, and Effect

The SAT does not just address a test taker's ability to find facts within a historical reading passage; it also evaluates a reader's ability to determine an author's viewpoint through the use of tone, message, and overall effect. This type of reading comprehension requires inference skills, deductive reasoning skills, the ability to draw logical conclusions, and overall critical thinking skills. Reading for factual information is straightforward. Reading for an author's tone, message, and overall effect is not. It's key to read carefully when asked test questions that address a test taker's ability to these writing devices. These are not questions that can be easily answered by quickly scanning for the right information.

Tone

An author's **tone** is the use of particular words, phrases, and writing style to convey an overall meaning. Tone expresses the author's attitude towards a particular topic. For example, a historical reading passage may begin like the following:

> The presidential election of 1960 ushered in a new era, a new Camelot, a new phase of forward thinking in U.S. politics that embraced brash action, unrest, and responded with admirable leadership.

From this opening statement, a reader can draw some conclusions about the author's attitude towards President John F. Kennedy. Furthermore, the reader can make additional, educated guesses about the state of the Union during the 1960 presidential election. By close reading, the test taker can determine that the repeated use of the word *new* and words such as *admirable leadership* indicate the author's tone of admiration regarding the President Kennedy's boldness. In addition, the author assesses that the era during President Kennedy's administration was problematic through the use of the words *brash action* and *unrest.* Therefore, if a test taker encountered a test question asking about the author's use of tone and their assessment of the Kennedy administration, the test taker should be able to identify an answer indicating admiration. Similarly, if asked about the state of the Union during the 1960s, a test taker should be able to correctly identify an answer indicating political unrest.

When identifying an author's tone, the following list of words may be helpful. This is not an inclusive list. Generally, parts of speech that indicate attitude will also indicate tone:

- Comical
- Angry
- Ambivalent
- Scary
- Lyrical
- Matter-of-fact
- Judgmental
- Sarcastic
- Malicious
- Objective
- Pessimistic
- Patronizing
- Gloomy
- Instructional
- Satirical
- Formal
- Casual

Message

An author's **message** is the same as the overall meaning of a passage. It is the main idea, or the main concept the author wishes to convey. An author's message may be stated outright, or it may be implied. Regardless, the test taker will need to use careful reading skills to identify an author's message or purpose.

Often, the message of a particular passage can be determined by thinking about why the author wrote the information. Many historical passages are written to inform and to teach readers established, factual information. However, many historical works are also written to convey biased ideas to readers. Gleaning bias from an author's message in a historical passage can be difficult, especially if the reader is presented with a variety of established facts as well. Readers tend to accept historical writing as factual. This is not always the case. Any discerning reader who has tackled historical information on topics such as United States political party agendas can attest that two or more works on the same topic may have completely different messages supporting or refuting the value of the identical policies.

Therefore, it is important to critically assess an author's message separate from factual information. One author, for example, may point to the rise of unorthodox political candidates in an election year based on the failures of the political party in office while another may point to the rise of the same candidates in the same election year based on the current party's successes. The historical facts of what has occurred leading up to an election year are not in refute. Labeling those facts as a failure or a success is a bias within an author's overall *message*, as is excluding factual information in order to further a particular point. In a standardized testing situation, a reader must be able to critically assess what the author is trying to say separate from the historical facts that surround their message.

Using the example of Lincoln's Gettysburg Address, a test question may ask the following:

> What is the message the author is trying to convey through this address?

Then they will ask the test taker to select an answer that best expresses Lincoln's message to his audience. Based on the options given, a test taker should be able to select the answer expressing the idea that Lincoln's audience should recognize the efforts of those who died in the war as a sacrifice to preserving human equality and self-government.

Effect

The **effect** an author wants to convey is when an author wants to impart a particular mood in their message. An author may want to challenge a reader's intellect, inspire imagination, or spur emotion. An author may present information to appeal to a physical, aesthetic, or transformational sense. Take the following text as an example:

> In 1963, Martin Luther King stated "I have a dream." The gathering at the Lincoln Memorial was the beginning of the Civil Rights movement and, with its reference to the Emancipation Proclamation, Dr. King's words electrified those who wanted freedom and equality while rising from hatred and slavery. It was the beginning of radical change.

The test taker may be asked about the effect this statement might have on King's audience. Through careful reading of the passage, the test taker should be able to choose an answer that best identifies an effect of grabbing the audience's attention. The historical facts are in place: King made the speech in 1963 at the Lincoln Memorial, kicked off the civil rights movement, and referenced the Emancipation Proclamation. The words *electrified* and *radical change* indicate the effect the author wants the reader to understand as a result of King's speech. In this historical passage, facts are facts. However, the author's message goes beyond the facts to indicate the effect the message had on the audience and, in addition, the effect the event should have on the reader.

When reading historical passages, the test taker should perform due diligence in their awareness of the test questions and answers up front. From there, the test taker should carefully, and critically, read all historical excerpts with an eye for detail, tone, message (biased or unbiased), and effect. Being able to synthesize these skills will result in success in a standardized testing situation.

Science Passages

The SAT includes at least two science passages that address the fundamental concepts of Earth science, biology, chemistry, and/or physics. While prior general knowledge of these subjects is helpful in determining correct test answers, the test taker's ability to comprehend the passages is key to success. When reading scientific excerpts, the test taker must be able to examine quantitative information, identify hypotheses, interpret data, and consider implications of the material they are presented with. It is helpful, at this point, to reference the above section on comprehending test questions prior to reading. The same rules apply: read questions and scan questions, along with their answers, prior to fully reading a passage. Be informed prior to approaching a scientific text. A test taker should know what they will be asked and how to apply their reading skills. In this section of the test, it is also likely that a test taker will encounter graphs and charts to assess their ability to interpret scientific data with an appropriate conclusion. This section will determine the skills necessary to address scientific data presented through identifying hypotheses, through reading and examining data, and through interpreting data representation passages.

Examine Hypotheses

When presented with fundamental, scientific concepts, it is important to read for understanding. The most basic skill in achieving this literacy is to understand the concept of hypothesis and moreover, to be able to identify it in a particular passage. A **hypothesis** is a proposed idea that needs further investigation in order to be proven true or false. While it can be considered an educated guess, a hypothesis goes more in depth in its attempt to explain something that is not currently accepted within scientific theory. It requires further experimentation and data gathering to test its validity and is subject to change, based on scientifically conducted test results. Being able to read a science passage and understand its main purpose, including any hypotheses, helps the test taker understand data-driven evidence. It helps the test taker to be able to correctly answer questions about the science excerpt they are asked to read.

When reading to identify a hypothesis, a test taker should ask, "What is the passage trying to establish? What is the passage's main idea? What evidence does the passage contain that either supports or refutes this idea?" Asking oneself these questions will help identify a hypothesis. Additionally, hypotheses are logical statements that are testable and use very precise language.

Review the following hypothesis example:

> Consuming excess sugar in the form of beverages has a greater impact on childhood obesity and subsequent weight gain than excessive sugar from food.

While this is likely a true statement, it is still only a conceptual idea in a text passage regarding how sugar consumption affects childhood obesity, unless the passage also contains tested data that either proves or disproves the statement. A test taker could expect the rest of the passage to cite data proving that children who drink empty calories gain more weight, and are more likely to be obese, than children who eat sugary snacks.

A hypothesis goes further in that, given its ability to be proven or disproven, it may result in further hypotheses that require extended research. For example, the hypothesis regarding sugar consumption

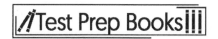

in drinks, after undergoing rigorous testing, may lead scientists to state another hypothesis such as the following:

> Consuming excess sugar in the form of beverages as opposed to food items is a habit found in mostly sedentary children.

This new, working hypothesis further focuses not just on the source of an excess of calories, but tries an "educated guess" that empty caloric intake has a direct, subsequent impact on physical behavior.

The data-driven chart below is similar to an illustration a test taker might see in relation to the hypothesis on sugar consumption in children:

Behaviors of Healthy and Unhealthy Kids

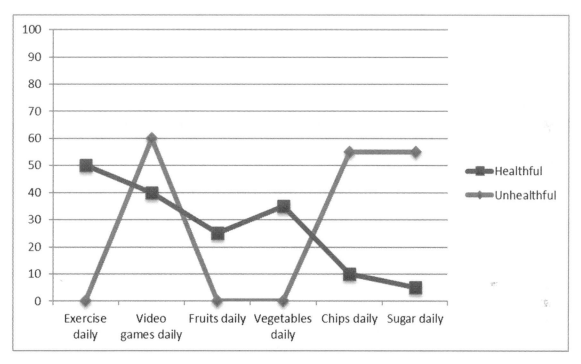

While this guide will address other data-driven passages a test taker could expect to see within a given science excerpt, note that the hypothesis regarding childhood sugar intake and rate of exercise has undergone scientific examination and yielded results that support its truth.

When reading a science passage to determine its hypothesis, a test taker should look for a concept that attempts to explain a phenomenon, is testable, logical, precisely worded, and yields data-driven results. The test taker should scan the presented passage for any word or data-driven clues that will help identify the hypothesis, and then be able to correctly answer test questions regarding the hypothesis based on their critical thinking skills.

Interpreting Data and Considering Implications

The Reading section of the SAT will contain one data-driven science passage that requires the test taker to examine evidence within a particular type of graphic. The test taker will then be required to interpret the data and answer questions demonstrating their ability to draw logical conclusions.

In general, there are two types of data: qualitative and quantitative. Science passages may contain both, but simply put, **quantitative** data is reflected numerically and qualitative is not. **Qualitative** data is based on its qualities. In other words, qualitative data tends to present information more in subjective generalities (for example, relating to size or appearance). Quantitative data is based on numerical findings such as percentages. Quantitative data will be described in numerical terms. While both types of data are valid, the test taker will more likely be faced with having to interpret quantitative data through one or more graphic(s), and then be required to answer questions regarding the numerical data. The section of this study guide briefly addresses how data may be displayed in line graphs, bar charts, circle graphs, and scatter plots. A test taker should take the time to learn the skills it takes to interpret quantitative data. An example of a line graph is as follows:

Cell Phone Use in Kiteville, 2000-2006

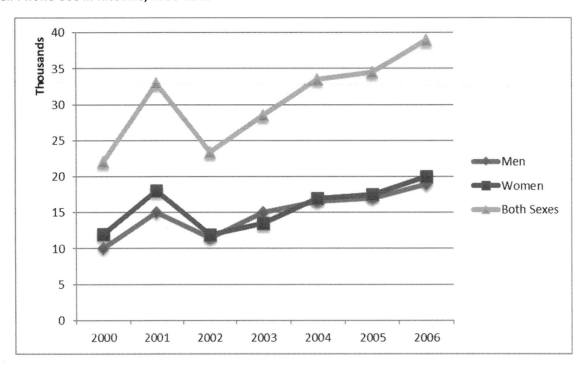

A **line graph** presents quantitative data on both horizontal (side to side) and vertical (up and down) axes. It requires the test taker to examine information across varying data points. When reading a line graph, a test taker should pay attention to any headings, as these indicate a title for the data it contains. In the above example, the test taker can anticipate the line graph contains numerical data regarding the use of cellphones during a certain time period. From there, a test taker should carefully read any outlying words or phrases that will help determine the meaning of data within the horizontal and vertical axes. In this example, the vertical axis displays the total number of people in increments of 5,000. Horizontally, the graph displays yearly markers, and the reader can assume the data presented accounts for a full calendar year. In addition, the line graph also uses different shapes to mark its data points. Some data points represent the number of men. Some data points represent the number of women, and a third type of data point represents the number of both sexes combined.

A test taker may be asked to read and interpret the graph's data, then answer questions about it. For example, the test may ask, *In which year did men seem to decrease cellphone use?* then require the test taker to select the correct answer. Similarly, the test taker may encounter a question such as *Which year*

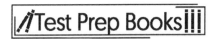

yielded the highest number of cellphone users overall? The test taker should be able to identify the correct answer as 2006.

A **bar graph** presents quantitative data through the use of lines or rectangles. The height and length of these lines or rectangles corresponds to the magnitude of the numerical data for that particular category or attribute. The data presented may represent information over time, showing shaded data over time or over other defined parameters. A bar graph will also utilize horizontal and vertical axes. An example of a bar graph is as follows:

Population Growth in Major U.S. Cities

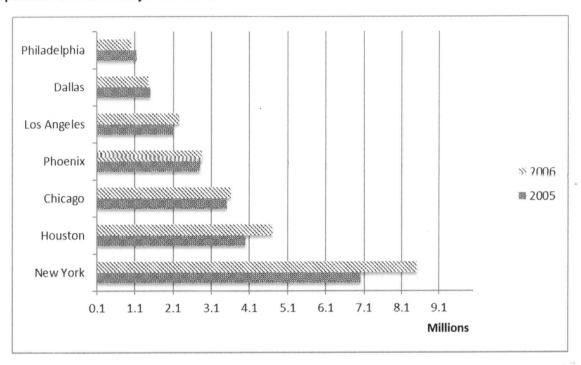

Reading the data in a bar graph is similar to the skills needed to read a line graph. The test taker should read and comprehend all heading information, as well as information provided along the horizontal and vertical axes. Note that the graph pertains to the population of some major U.S. cities. The "values" of these cities can be found along the left side of the graph, along the vertical axis. The population values can be found along the horizontal axes. Notice how the graph uses shaded bars to depict the change in population over time, as the heading indicates. Therefore, when the test taker is asked a question such as, *Which major U.S. city experienced the greatest amount of population growth during the depicted two year cycle,* the reader should be able to determine a correct answer of New York. It is important to pay particular attention to color, length, data points, and both axes, as well as any outlying header information in order to be able to answer graph-like test questions.

A **circle graph** (also sometimes referred to as a **pie chart**) presents quantitative data in the form of a circle. The same principles apply: the test taker should look for numerical data within the confines of the circle itself but also note any outlying information that may be included in a header, footer, or to the side of the circle. A circle graph will not depict horizontal or vertical axis information, but will instead rely on the reader's ability to visually take note of segmented circle pieces and apply information accordingly. An example of a circle graph is as follows:

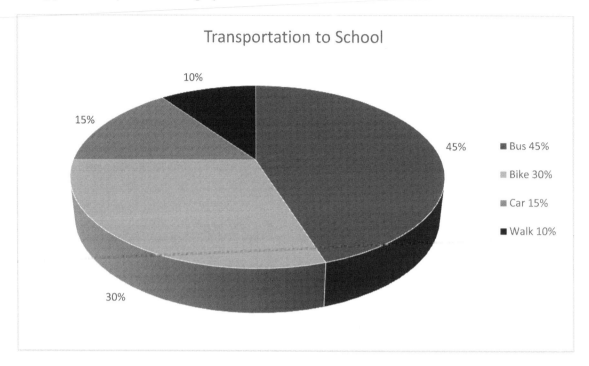

Notice the heading "Transportation to School." This should indicate to the test taker that the topic of the circle graph is how people traditionally get to school. To the right of the graph, the reader should comprehend that the data percentages contained within it directly correspond to the method of transportation. In this graph, the data is represented through the use shades and pattern. Each transportation method has its own shade. For example, if the test taker was then asked, *Which method of school transportation is most widely utilized,* the reader should be able to identify school bus as the correct answer.

Be wary of test questions that ask test takers to draw conclusions based on information that is not present. For example, it is not possible to determine, given the parameters of this circle graph, whether the population presented is of a particular gender or ethnic group. This graph does not represent data from a particular city or school district. It does not distinguish between student grade levels and, although the reader could infer that the typical student must be of driving age if cars are included, this is not necessarily the case. Elementary school students may rely on parents or others to drive them by personal methods. Therefore, do not read too much into data that is not presented. Only rely on the quantitative data that is presented in order to answer questions.

A **scatter plot** or **scatter diagram** is a graph that depicts quantitative data across plotted points. It will involve at least two sets of data. It will also involve horizontal and vertical axes.

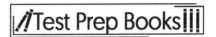

An example of a scatter plot is as follows:

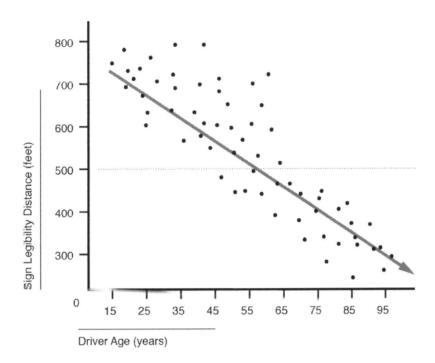

The skills needed to address a scatter plot are essentially the same as in other graph examples. Note any topic headings, as well as horizontal or vertical axis information. In the sample above, the reader can determine the data addresses a driver's ability to correctly and legibly read road signs as related to their age. Again, note the information that is absent. The test taker is not given the data to assess a time period, location, or driver gender. It simply requires the reader to note an approximate age to the ability to correctly identify road signs from a distance measured in feet. Notice that the overall graph also displays a trend. In this case, the data indicates a negative one and possibly supports the hypothesis that as a driver ages, their ability to correctly read a road sign at over 500 feet tends to decline over time. If the test taker were to be asked, *At what approximation in feet does a sixteen-year-old driver correctly see and read a street sign,* the answer would be the option closest to 700 feet.

Reading and examining scientific data in excerpts involves all of a reader's contextual reading, data interpretation, drawing logical conclusions based only on the information presented, and their application of critical thinking skills across a set of interpretive questions. Thorough comprehension and attention to detail is necessary to achieve test success.

Practice Questions

Questions 1–5 are based on the following passage from Common Sense *by Thomas Paine:*

MANKIND being originally equals in the order of creation, the equality could only be destroyed by some subsequent circumstance; the distinctions of rich, and poor, may in a great measure be accounted for, and that without having recourse to the harsh ill sounding names of oppression and avarice. Oppression is often the consequence, but seldom or never the means of riches; and though avarice will preserve a man from being necessitously poor, it generally makes him too timorous to be wealthy.

But there is another and greater distinction for which no truly natural or religious reason can be assigned, and that is, the distinction of men into KINGS and SUBJECTS. Male and female are the distinctions of nature, good and bad the distinctions of heaven; but how a race of men came into the world so exalted above the rest, and distinguished like some new species, is worth enquiring into, and whether they are the means of happiness or of misery to mankind.

In the early ages of the world, according to the scripture chronology, there were no kings; the consequence of which was there were no wars; it is the pride of kings which throw mankind into confusion Holland. without a king hath enjoyed more peace for this last century than any of the monarchical governments in Europe. Antiquity favors the same remark; for the quiet and rural lives of the first patriarchs hath a happy something in them, which vanishes away when we come to the history of Jewish royalty.

Government by kings was first introduced into the world by the Heathens, from whom the children of Israel copied the custom. It was the most prosperous invention the Devil ever set on foot for the promotion of idolatry. The Heathens paid divine honors to their deceased kings, and the Christian world hath improved on the plan by doing the same to their living ones. How impious is the title of sacred majesty applied to a worm, who in the midst of his splendor is crumbling into dust!

As the exalting one man so greatly above the rest cannot be justified on the equal rights of nature, so neither can it be defended on the authority of scripture; for the will of the Almighty, as declared by Gideon and the prophet Samuel, expressly disapproves of government by kings. All anti-monarchical parts of scripture have been very smoothly glossed over in monarchical governments, but they undoubtedly merit the attention of countries, which have their governments yet to form. "Render unto Caesar the things which are Caesar's" is the scripture doctrine of courts, yet it is no support of monarchical government, for the Jews at that time were without a king, and in a state of vassalage to the Romans.

Near three thousand years passed away from the Mosaic account of the creation, till the Jews under a national delusion requested a king. Till then their form of government (except in extraordinary cases, where the Almighty interposed) was a kind of republic administered by a judge and the elders of the tribes. Kings they had none, and it was held sinful to acknowledge any being under that title but the Lord of Hosts. And when a man seriously reflects on the idolatrous homage which is paid to the persons of Kings,

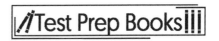

he need not wonder, that the Almighty ever jealous of his honor, should disapprove of a form of government which so impiously invades the prerogative of heaven.

1. According to passage, what role does avarice, or greed, play in poverty?
    a. It can make a man very wealthy.
    b. It is the consequence of wealth.
    c. Avarice can prevent a man from being poor, but too fearful to be very wealthy.
    d. Avarice is what drives a person to be very wealthy.

2. Of these distinctions, which does the author believe to be beyond natural or religious reason?
    a. Good and bad
    b. Male and female
    c. Human and animal
    d. King and subjects

3. According to the passage, what are the Heathens responsible for?
    a. Government by kings
    b. Quiet and rural lives of patriarchs
    c. Paying divine honors to their living kings
    d. Equal rights of nature

4. Which of the following best states Paine's rationale for the denouncement of monarchy?
    a. It is against the laws of nature.
    b. It is against the equal rights of nature and is denounced in scripture.
    c. Despite scripture, a monarchal government is unlawful.
    d. Neither the law nor scripture denounce monarchy.

5. Based on the passage, what is the best definition of the word *idolatrous*?
    a. Worshipping heroes
    b. Being deceitful
    c. Sinfulness
    d. Engaging in illegal activities

Answer Explanations

1. C: In the first paragraph, it is stated that avarice can prevent a man from being necessitously poor, but too timorous, or fearful, to achieve real wealth. According to the passage, avarice does not tend to make a person very wealthy. The passage states that oppression, not avarice, is the consequence of wealth. The passage does not state that avarice drives a person's desire to be wealthy.

2. D: Paine believes that the distinction that is beyond a natural or religious reason is between king and subjects. He states that the distinction between good and bad is made in heaven. The distinction between male and female is natural. He does not mention anything about the distinction between humans and animals.

3. A: The passage states that the Heathens were the first to introduce government by kings into the world. The quiet lives of patriarchs came before the Heathens introduced this type of government. It was Christians, not Heathens, who paid divine honors to living kings. Heathens honored deceased kings. Equal rights of nature are mentioned in the paragraph, but not in relation to the Heathens.

4. B: Paine asserts that a monarchy is against the equal rights of nature and cites several parts of scripture that also denounce it. He doesn't say it is against the laws of nature. Because he uses scripture to further his argument, it is not despite scripture that he denounces the monarchy. Paine addresses the law by saying the courts also do not support a monarchical government.

5. A: To be *idolatrous* is to worship idols or heroes, in this case, kings. It is not defined as being deceitful. While idolatry is considered a sin, it is an example of a sin, not a synonym for it. Idolatry may have been considered illegal in some cultures, but it is not a definition for the term.

Writing and Language

The Writing and Language section of the SAT presents passages with deliberate errors; the section is designed to assess the test taker's command of standard conventions of English Language and his or her ability to edit errors appropriately. While questions will ask test takers to select more effective alternatives to given sentences and revise and edit errors, they will not require test takers to identify specific grammatical rules.

The section includes four passages, each of approximately 400-450 words in length. There is one passage dedicated to each of the following four categories: careers, science, humanities, and social studies. Careers passages present trends or debates in various industries like technology, communications, or healthcare. Science passages are drawn from topics in earth science, chemistry, biology, or physics. Humanities passages may focus on any number of topics such as arts, literature, dance, or music. Social studies passages may pull topics from a variety of areas including history, sociology, psychology, anthropology, and political science. All of the information needed to answer the questions is provided within the passages and test takers do not need to have prior knowledge of the passages' content to obtain the correct answers.

Test takers are given 35 minutes to read the four passages and then answer a total of 44 multiple-choice questions—eleven per passage. In addition to surveying the test taker's ability to comprehend and revise passages about various topics, the section tests one's skill in understanding various writing styles. The test taker will encounter at least one passage written in one of three different writing styles: narrative, argumentative, and explanatory or informative.

In addition, test takers can expect to interpret graphs, tables, charts, or other graphics that, in some way, support factual content or an opinion presented in the graphic's passage. After reading the passage and reviewing the graphic, test takers may be asked to correct an error in the passage that contradicts information presented in the graphic. Alternatively, they may need to use information gleaned from the graphic to select a more specific description for part or all of the passage in place of the vague description originally provided. Essentially, the passage may misinterpret the graphic or the graphic may otherwise illuminate important data or facts that can strengthen the argument or explanation in the passage; test takers will be required to identify these findings.

Passages and the questions pertaining to them usually appear side by side on the test, with the passage on the left side of the page and the questions on the right. Some or all passages may span multiple pages, so test takers should ensure they have read the entire passage before answering questions that pertain to the passage in its entirety. In contrast, some test takers find it useful to answer questions addressing specific points or details as that information is encountered in the passage; question numbers embedded within the passage should be used to guide the test taker to the section of the passage for which a question pertains. Test takers should then carefully read the underlined text in the passage following the number and determine if errors are present.

The related question will then ask test takers to consider the underlined sentence or section and the provided answer choices. Test takers must select the answer choice that offers that best revision or most effective replacement for the underlined section. If the original sentence seems correct as is, test takers should select Choice A—NO CHANGE—which represents that the initial sentence was free from errors and should remain unchanged.

Underlined text will not necessarily follow all of the embedded boxed numbers in the passage. In these situations, test takers should still use the boxed number to guide their attention to the pertinent question, and then the appropriate question will inform them on what to do. Some sentences or paragraphs may be numbered, and test takers will have to decide how that text should be ordered or where a particular sentence or paragraph should be inserted.

There are two principle question types on the Writing and Language Section. For the Expression of Ideas questions, test takers will select answers that improve the text's organization, topic development, and the effectiveness of the language used. The scope of these questions assesses one's understanding of text structure and organization, ability to analyze evidence and arguments and strengthen them with designated improvements, and his or her specific skills in comprehending passages addressing topics in science, history, and the humanities. Standard English Convention questions address one's skill in identifying and correcting errors in sentence structure, word usage, grammar, and punctuation. These questions evaluate the test taker's skill in determining the correct meaning of words in context, selecting precise words, and revising text to be more concise, coherent, and consistent in tone and style.

While the vast majority of questions in the Writing and Language section are comprised of these two question types in roughly equal proportions, as mentioned, a small number of questions will also assess the test taker's ability to interpret graph, tables, and charts. For this reason, in addition to studying grammatical rules, vocabulary, and word usage and developing writing skills and an eye and ear for issues with organization and flow, test takers should practice reading and interpreting tables, graphs, and charts. As such, the sections that follow provide a review of material pertinent to success in this section, covering topics on English language conventions, text structure and readability, syntax and grammar, analyzing opinions and facts, and interpreting graphics.

Expression of Ideas

This Writing and Language section is about *how* the information is communicated rather than the subject matter itself. The good news is there isn't any writing! Instead, it's like being an editor helping the writer find the best ways to express their ideas. Things to consider include: how well a topic is developed, how accurately facts are presented, whether the writing flows logically and cohesively, and how effectively the writer uses language. This can seem like a lot to remember, but these concepts are the same ones taught way back in elementary school.

One last thing to remember while going through this guide is not to be intimidated by the terminology. Phrases like "pronoun-antecedent agreement" and "possessive determiners" can sound confusing and complicated, but the ideas are often quite simple and easy to understand. Though proper terminology is used to explain the rules and guidelines, the SAT Writing and Language Test is not a technical grammar test.

Organization

Good writing is not merely a random collection of sentences. No matter how well written, sentences must relate and coordinate appropriately to one another. If not, the writing seems random, haphazard, and disorganized. Therefore, good writing must be **organized** (where each sentence fits a larger context and relates to the sentences around it).

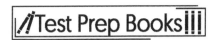

Transition Words

The writer should act as a guide, showing the reader how all the sentences fit together. Consider this example:

> Seat belts save more lives than any other automobile safety feature. Many studies show that airbags save lives as well. Not all cars have airbags. Many older cars don't. Air bags aren't entirely reliable. Studies show that in 15% of accidents, airbags don't deploy as designed. Seat belt malfunctions are extremely rare.

There's nothing wrong with any of these sentences individually, but together they're disjointed and difficult to follow. The best way for the writer to communicate information is through the use of **transition words**. Here are examples of transition words and phrases that tie sentences together, enabling a more natural flow:

- To show causality: *as a result, therefore*, and *consequently*
- To compare and contrast: *however, but,* and *on the other hand*
- To introduce examples: *for instance, namely*, and *including*
- To show order of importance: *foremost, primarily, secondly*, and *lastly*

The above is not a complete list of transitions. There are many more that can be used; however, most fit into these or similar categories. The important point is that the words should clearly show the relationship between sentences, supporting information, and the main idea.

Here is an update to the previous example using transition words. These changes make it easier to read and bring clarity to the writer's points:

> Seat belts save more lives than any other automobile safety feature. Many studies show that airbags save lives as well. However, not all cars have airbags. For instance, some older cars don't. Furthermore, air bags aren't entirely reliable. For example, studies show that in 15% of accidents, airbags don't deploy as designed. But, on the other hand, seat belt malfunctions are extremely rare.

Also be prepared to analyze whether the writer is using the best transition word or phrase for the situation. Take this sentence for example: "As a result, seat belt malfunctions are extremely rare." This sentence doesn't make sense in the context above because the writer is trying to show the **contrast** between seat belts and airbags, not the causality.

Logical Sequence

Even if the writer includes plenty of information to support their point, the writing is only effective when the information is in a logical order. **Logical sequencing** is really just common sense, but it's also an important writing technique. First, the writer should introduce the main idea, whether for a paragraph, a section, or the entire piece. Second, they should present evidence to support the main idea by using transitional language. This shows the reader how the information relates to the main idea and to the sentences around it. The writer should then take time to interpret the information, making sure necessary connections are obvious to the reader. Finally, the writer can summarize the information in a closing section.

Although most writing follows this pattern, it isn't a set rule. Sometimes writers change the order for effect. For example, the writer can begin with a surprising piece of supporting information to grab the reader's attention, and then transition to the main idea. Thus, if a passage doesn't follow the logical order, don't immediately assume it's wrong. However, most writing usually settles into a logical sequence after a nontraditional beginning.

Focus

Good writing stays **focused** and on topic. During the test, determine the main idea for each passage and then look for times when the writer strays from the point they're trying to make. Let's go back to the seat belt example. If the writer suddenly begins talking about how well airbags, crumple zones, or other safety features work to save lives, they might be losing focus from the topic of "safety belts."

Focus can also refer to individual sentences. Sometimes the writer does address the main topic, but in a confusing way. For example:

> Thanks to seat belt usage, survival in serious car accidents has shown a consistently steady increase since the development of the retractable seat belt in the 1950s.

This statement is definitely on topic, but it's not easy to follow. A simpler, more focused version of this sentence might look like this:

> Seat belts have consistently prevented car fatalities since the 1950s.

Providing **adequate information** is another aspect of focused writing. Statements like "seat belts are important" and "many people drive cars" are true, but they're so general that they don't contribute much to the writer's case. When reading a passage, watch for these kinds of unfocused statements.

Introductions and Conclusions

Examining the writer's strategies for introductions and conclusions puts the reader in the right mindset to interpret the rest of the passage. Look for methods the writer might use for introductions such as:

- Stating the main point immediately, followed by outlining how the rest of the piece supports this claim.

- Establishing important, smaller pieces of the main idea first, and then grouping these points into a case for the main idea.

- Opening with a quotation, anecdote, question, seeming paradox, or other piece of interesting information, and then using it to lead to the main point.

Whatever method the writer chooses, the **introduction** should make their intention clear, establish their voice as a credible one, and encourage a person to continue reading.

Conclusions tend to follow a similar pattern. In them, the writer restates their main idea a final time, often after summarizing the smaller pieces of that idea. If the introduction uses a quote or anecdote to grab the reader's attention, the conclusion often makes reference to it again. Whatever way the writer chooses to arrange the conclusion, the final restatement of the main idea should be clear and simple for the reader to interpret.

Finally, conclusions shouldn't introduce any new information.

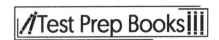

Precision

People often think of **precision** in terms of math, but precise word choice is another key to successful writing. Since language itself is imprecise, it's important for the writer to find the exact word or words to convey the full, intended meaning of a given situation. For example:

> The number of deaths has gone down since seat belt laws started.

There are several problems with this sentence. First, the word *deaths* is too general. From the context, it's assumed that the writer is referring only to *deaths* caused by car accidents. However, without clarification, the sentence lacks impact and is probably untrue. The phrase "gone down" might be accurate, but a more precise word could provide more information and greater accuracy. Did the numbers show a slow and steady decrease of highway fatalities or a sudden drop? If the latter is true, the writer is missing a chance to make their point more dramatically. Instead of "gone down" they could substitute *plummeted*, *fallen drastically*, or *rapidly diminished* to bring the information to life. Also, the phrase "seat belt laws" is unclear. Does it refer to laws requiring cars to include seat belts or to laws requiring drivers and passengers to use them? Finally, *started* is not a strong verb. Words like *enacted* or *adopted* are more direct and make the content more real. When put together, these changes create a far more powerful sentence:

> The number of highway fatalities has plummeted since laws requiring seat belt usage were enacted.

However, it's important to note that precise word choice can sometimes be taken too far. If the writer of the sentence above takes precision to an extreme, it might result in the following:

> The incidence of high-speed, automobile accident related fatalities has decreased 75% and continued to remain at historical lows since the initial set of federal legislations requiring seat belt use were enacted in 1992.

This sentence is extremely precise, but it takes so long to achieve that precision that it suffers from a lack of clarity. Precise writing is about finding the right balance between information and flow. This is also an issue of **conciseness** (discussed in the next section).

The last thing to consider with precision is a word choice that's not only unclear or uninteresting, but also confusing or misleading. For example:

> The number of highway fatalities has become hugely lower since laws requiring seat belt use were enacted.

In this case, the reader might be confused by the word *hugely*. Huge means large, but here the writer uses *hugely* to describe something small. Though most readers can decipher this, doing so disconnects them from the flow of the writing and makes the writer's point less effective.

On the test, there can be questions asking for alternatives to the writer's word choice. In answering these questions, always consider the context and look for a balance between precision and flow.

Conciseness

"Less is more" is a good rule to follow when writing a sentence. Unfortunately, writers often include extra words and phrases that seem necessary at the time, but add nothing to the main idea. This confuses the reader and creates unnecessary repetition. Writing that lacks **conciseness** is usually guilty of excessive wordiness and redundant phrases. Here's an example containing both of these issues:

> When legislators decided to begin creating legislation making it mandatory for automobile drivers and passengers to make use of seat belts while in cars, a large number of them made those laws for reasons that were political reasons.

There are several empty or "fluff" words here that take up too much space. These can be eliminated while still maintaining the writer's meaning. For example:

- "decided to begin" could be shortened to "began"
- "making it mandatory for" could be shortened to "requiring"
- "make use of" could be shortened to "use"
- "a large number" could be shortened to "many"

In addition, there are several examples of redundancy that can be eliminated:

- "legislators decided to begin creating legislation" and "made those laws"
- "automobile drivers and passengers" and "while in cars"
- "reasons that were political reasons"

These changes are incorporated as follows:

> When legislators began requiring drivers and passengers to use seat belts, many of them did so for political reasons.

There are many examples of redundant phrases, such as "add an additional," "complete and total," "time schedule," and "transportation vehicle." If asked to identify a redundant phrase on the test, look for words that are close together with the same (or similar) meanings.

Proposition

The **proposition** (also called the **claim** since it can be true or false) is a clear statement of the point or idea the writer is trying to make. The length or format of a proposition can vary, but it often takes the form of a **topic sentence**. A good topic sentence is:

- Clear: does not weave a complicated web of words for the reader to decode or unwrap

- Concise: presents only the information needed to make the claim and doesn't clutter up the statement with unnecessary details

- Precise: clarifies the exact point the writer wants to make and doesn't use broad, overreaching statements

Look at the following example:

> The civil rights movement, from its genesis in the Emancipation Proclamation to its current struggles with de facto discrimination, has changed the face of the United States more than any other factor in its history.

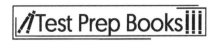

Is the statement clear? Yes, the statement is fairly clear, although other words can be substituted for "genesis" and "de facto" to make it easier to understand.

Is the statement concise? No, the statement is not concise. Details about the Emancipation Proclamation and the current state of the movement are unnecessary for a topic sentence. Those details should be saved for the body of the text.

Is the statement precise? No, the statement is not precise. What exactly does the writer mean by "changed the face of the United States"? The writer should be more specific about the effects of the movement. Also, suggesting that something has a greater impact than anything else in U.S. history is far too ambitious a statement to make.

A better version might look like this:

> The civil rights movement has greatly increased the career opportunities available for Black Americans.

The unnecessary language and details are removed, and the claim can now be measured and supported.

Support

Once the main idea or proposition is stated, the writer attempts to prove or **support** the claim with text evidence and supporting details.

Take for example the sentence, "Seat belts save lives." Though most people can't argue with this statement, its impact on the reader is much greater when supported by additional content. The writer can support this idea by:

- Providing statistics on the rate of highway fatalities alongside statistics for estimated seat belt usage.

- Explaining the science behind a car accident and what happens to a passenger who doesn't use a seat belt.

- Offering anecdotal evidence or true stories from reliable sources about how seat belts prevent fatal injuries in car crashes.

However, using only one form of supporting evidence is not nearly as effective as using a variety to support a claim. Presenting only a list of statistics can be boring to the reader, but providing a true story that's both interesting and humanizing helps. In addition, one example isn't always enough to prove the writer's larger point, so combining it with other examples is extremely effective for the writing. Thus, when reading a passage, don't just look for a single form of supporting evidence.

Another key aspect of supporting evidence is a **reliable source**. Does the writer include the source of the information? If so, is the source well known and trustworthy? Is there a potential for bias? For example, a seat belt study done by a seat belt manufacturer may have its own agenda to promote.

Effective Language Use

Language can be analyzed in a variety of ways. But one of the primary ways is its effectiveness in communicating and especially convincing others.

Rhetoric is a literary technique used to make the writing (or speaking) more effective or persuasive. Rhetoric makes use of other effective language devices such as irony, metaphors, allusion, and repetition. An example of the rhetorical use of repetition would be: "Let go, I say, let go!!!".

Figures of Speech

A **figure of speech** (sometimes called an **idiom**) is a rhetorical device. It's a phrase that's not intended to be taken literally.

When the writer uses a figure of speech, their intention must be clear if it's to be used effectively. Some phrases can be interpreted in a number of ways, causing confusion for the reader. In the SAT Writing and Language Test, questions may ask for an alternative to a problematic word or phrase. Look for clues to the writer's true intention to determine the best replacement. Likewise, some figures of speech may seem out of place in a more formal piece of writing. To show this, here is the previous seat belt example but with one slight change:

> Seat belts save more lives than any other automobile safety feature. Many studies show that airbags save lives as well. However, not all cars have airbags. For instance, some older cars don't. In addition, air bags aren't entirely reliable. For example, studies show that in 15% of accidents, airbags don't deploy as designed. But, on the other hand, seat belt malfunctions happen once in a blue moon.

Most people know that "once in a blue moon" refers to something that rarely happens. However, because the rest of the paragraph is straightforward and direct, using this figurative phrase distracts the reader. In this example, the earlier version is much more effective.

Now it's important to take a moment and review the meaning of the word *literally*. This is because it's one of the most misunderstood and misused words in the English language. **Literally** means that something is exactly what it says it is, and there can be no interpretation or exaggeration. Unfortunately, *literally* is often used for emphasis as in the following example:

> This morning, I literally couldn't get out of bed.

This sentence meant to say that the person was extremely tired and wasn't able to get up. However, the sentence can't *literally* be true unless that person was tied down to the bed, paralyzed, or affected by a strange situation that the writer (most likely) didn't intend. Here's another example:

> I literally died laughing.

The writer tried to say that something was very funny. However, unless they're writing this from beyond the grave, it can't *literally* be true.

Rhetorical Fallacies

A **rhetorical fallacy** is an argument that doesn't make sense. It usually involves distracting the reader from the issue at hand in some way. There are many kinds of rhetorical fallacies. Here are just a few, along with examples of each:

- **Ad Hominem:** Makes an irrelevant attack against the person making the claim, rather than addressing the claim itself. For example, Senator Wilson opposed the new seat belt legislation, but should we really listen to someone who's been divorced four times?

- **Exaggeration:** Represents an idea or person in an obviously excessive manner. For example, Senator Wilson opposed the new seat belt legislation. Maybe she thinks if more people die in car accidents, it will help with overpopulation.

- **Stereotyping** (or **Categorical Claim**): Claims that all people of a certain group are the same in some way. For example, Senator Wilson still opposes the new seat belt legislation. You know women can never admit when they're wrong.

When examining a possible rhetorical fallacy, carefully consider the point the writer is trying to make and if the argument directly relates to that point. If something feels wrong, there's a good chance that a fallacy is at play. The SAT Writing and Language section doesn't expect the fallacy to be named using specific terms like those above. However, questions can include identifying why something is a fallacy or suggesting a sounder argument.

Style, Tone, and Mood

Style, tone, and mood are often thought to be the same thing. Though they're closely related, there are important differences to keep in mind. The easiest way to do this is to remember that style "creates and affects" tone and mood. More specifically, **style** is *how the writer uses words* to create the desired tone and mood for their writing.

Style

Style can include any number of technical writing choices, and some may have to be analyzed on the test. A few examples of style choices include:

- Sentence Construction: When presenting facts, does the writer use shorter sentences to create a quicker sense of the supporting evidence, or do they use longer sentences to elaborate and explain the information?

- Technical Language: Does the writer use jargon to demonstrate their expertise in the subject, or do they use ordinary language to help the reader understand things in simple terms?

- Formal Language: Does the writer refrain from using contractions such as *won't* or *can't* to create a more formal tone, or do they use a colloquial, conversational style to connect to the reader?

- Formatting: Does the writer use a series of shorter paragraphs to help the reader follow a line of argument, or do they use longer paragraphs to examine an issue in great detail and demonstrate their knowledge of the topic?

On the test, examine the writer's style and how their writing choices affect the way the passage comes across.

Tone

As mentioned, **tone** refers to the writer's attitude toward the subject matter. Tone is usually explained in terms of a work of fiction. For example, the tone conveys how the writer feels about their characters and the situations in which they're involved. Nonfiction writing is sometimes thought to have no tone at all, but this is incorrect.

A lot of nonfiction writing has a neutral tone, which is an extremely important tone for the writer to take. A neutral tone demonstrates that the writer is presenting a topic impartially and letting the information speak for itself. On the other hand, nonfiction writing can be just as effective and appropriate if the tone isn't neutral. For instance, take the previous examples involving seat belt use. In them, the writer mostly chooses to retain a neutral tone when presenting information. If the writer would instead include their own personal experience of losing a friend or family member in a car accident, the tone would change dramatically. The tone would no longer be neutral. Now it would show that the writer has a personal stake in the content, allowing them to interpret the information in a different way. When analyzing tone, consider what the writer is trying to achieve in the passage, and how they *create* the tone using style.

Mood

Mood refers to the feelings and atmosphere that the writer's words create for the reader. Like tone, many nonfiction pieces can have a neutral mood. To return to the previous example, if the writer would choose to include information about a person they know being killed in a car accident, the passage would suddenly carry an emotional component that is absent in the previous examples. Depending on how they present the information, the writer can create a sad, angry, or even hopeful mood. When analyzing the mood, consider what the writer wants to accomplish and whether the best choice was made to achieve that end.

Consistency

Whatever style, tone, and mood the writer uses, good writing should remain **consistent** throughout. If the writer chooses to include the tragic, personal experience above, it would affect the style, tone, and mood of the entire piece. It would seem out of place for such an example to be used in the middle of a neutral, measured, and analytical piece. To adjust the rest of the piece, the writer needs to make additional choices to remain consistent. For example, the writer might decide to use the word *tragedy* in place of the more neutral *fatality*, or they could describe a series of car-related deaths as an *epidemic*. Adverbs and adjectives such as *devastating* or *horribly* could be included to maintain this consistent attitude toward the content. When analyzing writing, look for sudden shifts in style, tone, and mood, and consider whether the writer would be wiser to maintain the prevailing strategy.

Syntax

Syntax is the order of words in a sentence. While most of the writing on the test has proper syntax, there may be questions on ways to vary the syntax for effectiveness. One of the easiest writing mistakes to spot is **repetitive sentence structure**. For example:

> Seat belts are important. They save lives. People don't like to use them. We have to pass seat belt laws. Then more people will wear seat belts. More lives will be saved.

What's the first thing that comes to mind when reading this example? The short, choppy, and repetitive sentences! In fact, most people notice this syntax issue more than the content itself. By combining some sentences and changing the syntax of others, the writer can create a more effective writing passage:

> Seat belts are important because they save lives. Since people don't like to use seat belts, though, more laws requiring their usage need to be passed. Only then will more people wear them and only then will more lives be saved.

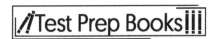

Many rhetorical devices can be used to vary syntax (more than can possibly be named here). These often have intimidating names like anadiplosis, metastasis, and paremptosis. The test questions don't ask for definitions of these tricky techniques, but they can ask how the writer plays with the words and what effect that has on the writing. For example, **anadiplosis** is when the last word (or phrase) from a sentence is used to begin the next sentence:

Cars are driven by people. People cause accidents. Accidents cost taxpayers money.

The test doesn't ask for this technique by name, but be prepared to recognize what the writer is doing and why they're using the technique in this situation. In this example, the writer is probably using anadiplosis to demonstrate causation.

Quantitative Information

As mentioned, some passages in the test contains **infographics** such as charts, tables, or graphs. In these cases, interpret the information presented and determine how well it supports the claims made in the text. For example, if the writer makes a case that seat belts save more lives than other automobile safety measures, they might want to include a graph (like the one below) showing the number of lives saved by seat belts versus those saved by air bags.

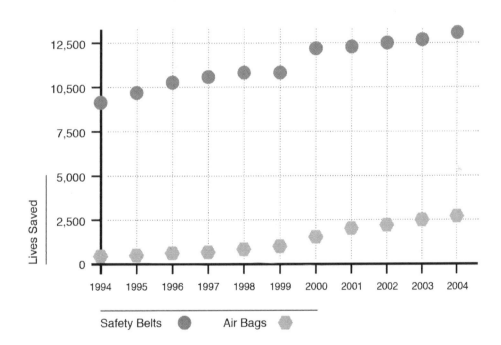

Based on data from the National Highway Traffic Safety Administration

If the graph clearly shows a higher number of lives are saved by seat belts, then it's effective. However, if the graph shows air bags save more lives than seat belts, then it doesn't support the writer's case.

Finally, graphs should be easy to understand. Their information should immediately be clear to the reader at a glance. Here are some basic things to keep in mind when interpreting infographics:

- In a **bar graph**, higher bars represent larger numbers. Lower bars represent smaller numbers.

- **Line graphs** often show trends over time. Points that are higher represent larger numbers than points that are lower. A line that consistently ascends from left to right shows a steady increase over time. A line that consistently descends from left to right shows a steady decrease over time. A line that bounces up and down represents instability or inconsistency in the trend. When interpreting a line graph, determine the point the writer is trying to make, and then see if the graph supports that point.

- **Pie charts** are used to show proportions or percentages of a whole but are less effective in showing change over time.

- **Tables** present information in numerical form, not as graphics. When interpreting a table, make sure to look for patterns in the numbers.

There can also be timelines, illustrations, or maps on the test. When interpreting these, keep in mind the writer's intentions and determine whether or not the graphic supports the case.

Standard English Conventions

Types of Sentences

There isn't an overabundance of absolutes in grammar, but here is one: every sentence in the English language falls into one of four categories.

- **Declarative:** a simple statement that ends with a period

    The price of milk per gallon is the same as the price of gasoline.

- **Imperative:** a command, instruction, or request that ends with a period

    Buy milk when you stop to fill up your car with gas.

- **Interrogative:** a question that ends with a question mark

    Will you buy the milk?

- **Exclamatory:** a statement or command that expresses emotions like anger, urgency, or surprise and ends with an exclamation mark

    Buy the milk now!

Declarative sentences are the most common type, probably because they are comprised of the most general content, without any of the bells and whistles that the other three types contain. They are, simply, declarations or statements of any degree of seriousness, importance, or information.

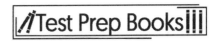

Imperative sentences often seem to be missing a subject. The subject is there, though; it is just not visible or audible because it is *implied*. Look at the imperative example sentence.

>   Buy the milk when you fill up your car with gas.

You is the implied subject, the one to whom the command is issued. This is sometimes called **the understood you** because it is understood that *you* is the subject of the sentence.

Interrogative sentences—those that ask questions—are defined as such from the idea of the word **interrogation**, the action of questions being asked of suspects by investigators. Although that is serious business, interrogative sentences apply to all kinds of questions.

To exclaim is at the root of **exclamatory** sentences. These are made with strong emotions behind them. The only technical difference between a declarative or imperative sentence and an exclamatory one is the exclamation mark at the end. The example declarative and imperative sentences can both become an exclamatory one simply by putting an exclamation mark at the end of the sentences.

>   The price of milk per gallon is the same as the price of gasoline!
>   Buy milk when you stop to fill up your car with gas!

After all, someone might be really excited by the price of gas or milk, or they could be mad at the person that will be buying the milk! However, as stated before, exclamation marks in abundance defeat their own purpose! After a while, they begin to cause fatigue! When used only for their intended purpose, they can have their expected and desired effect.

Parts of Speech

Nouns

A **noun** is a person, place, thing ,or idea. All nouns fit into one of two types, common or proper.

A **common noun** is a word that identifies any of a class of people, places, or things. Examples include numbers, objects, animals, feelings, concepts, qualities, and actions. *A, an,* or *the* usually precedes the common noun. These parts of speech are called *articles*. Here are some examples of sentences using nouns preceded by articles.

>   *A* building is under construction.
>   *The* girl would like to move to *the* city.

A **proper noun** (also called a **proper name**) is used for the specific name of an individual person, place, or organization. The first letter in a proper noun is capitalized. "My name is *Mary*." "I work for *Walmart*."

Nouns sometimes serve as adjectives (which themselves describe nouns), such as "hockey player" and "state government."

An **abstract noun** is an idea, state, or quality. It is something that can't be touched, such as happiness, courage, evil, or humor.

A **concrete noun** is something that can be experienced through the senses (touch, taste, hear, smell, see). Examples of concrete nouns are birds, skateboard, pie, and car.

A **collective noun** refers to a collection of people, places, or things that act as one. Examples of collective nouns are as follows: team, class, jury, family, audience, and flock.

Pronouns

A word used in place of a noun is known as a **pronoun.** Pronouns are words like *I, mine, hers,* and *us.*

Pronouns can be split into different classifications (seen below) which make them easier to learn; however, it's not important to memorize the classifications.

- Personal pronouns: refer to people
  - First person: we, I, our, mine
  - Second person: you, yours
  - Third person: he, them
- Possessive pronouns: demonstrate ownership (mine, his, hers, its, ours, theirs, yours)
- Interrogative pronouns: ask questions (what, which, who, whom, whose)
- Relative pronouns: include the five interrogative pronouns and others that are relative (whoever, whomever, that, when, where)
- Demonstrative pronouns: replace something specific (this, that, those, these)
- Reciprocal pronouns: indicate something was done or given in return (each other, one another)
- Indefinite pronouns: have a nonspecific status (anybody, whoever, someone, everybody, somebody)

Indefinite pronouns such as *anybody, whoever, someone, everybody,* and *somebody* command a singular verb form, but others such as *all, none,* and *some* could require a singular or plural verb form.

Antecedents

An **antecedent** is the noun to which a pronoun refers; it needs to be written or spoken before the pronoun is used. For many pronouns, antecedents are imperative for clarity. In particular, many of the personal, possessive, and demonstrative pronouns need antecedents. Otherwise, it would be unclear who or what someone is referring to when they use a pronoun like *he* or *this.*

Pronoun reference means that the pronoun should refer clearly to one, clear, unmistakable noun (the antecedent).

Pronoun-antecedent agreement refers to the need for the antecedent and the corresponding pronoun to agree in gender, person, and number. Here are some examples:

The *kidneys* (plural antecedent) are part of the urinary system. *They* (plural pronoun) serve several roles.

The kidneys are part of the *urinary system* (singular antecedent). *It* (singular pronoun) is also known as the renal system.

Pronoun Cases

The subjective pronouns —*I, you, he/she/it, we, they,* and *who*—are the subjects of the sentence.

Example: *They* have a new house.

The objective pronouns—*me, you* (*singular*), *him/her, us, them,* and *whom*—are used when something is being done for or given to someone; they are objects of the action.

Example: The teacher has an apple for *us*.

The possessive pronouns—*mine, my, your, yours, his, hers, its, their, theirs, our,* and *ours*—are used to denote that something (or someone) belongs to someone (or something).

Example: It's *their* chocolate cake.
Even Better Example: It's *my* chocolate cake!

One of the greatest challenges and worst abuses of pronouns concerns *who* and *whom*. Just knowing the following rule can eliminate confusion. *Who* is a subjective-case pronoun used only as a subject or subject complement. *Whom* is only objective-case and, therefore, the object of the verb or preposition.

Who is going to the concert?

You are going to the concert with *whom*?

Hint. When using *who* or *whom*, think of whether someone would say *he* or *him*. If the answer is *he*, use *who*. If the answer is *him*, use *whom*. This trick is easy to remember because *he* and *who* both end in vowels, and *him* and *whom* both end in the letter *M*.

Verbs

A **verb** is the part of speech that describes an action, state of being, or occurrence.

A verb forms the main part of a predicate of a sentence. This means that the verb explains what the noun (which will be discussed shortly) is doing. A simple example is *time flies*. The verb *flies* explains what the action of the noun, *time*, is doing. This example is a *main* verb.

Helping (auxiliary) verbs are words like *have, do, be, can, may, should, must,* and *will*. "I *should* go to the store." Helping verbs assist main verbs in expressing tense, ability, possibility, permission, or obligation.

Particles are minor function words like *not, in, out, up,* or *down* that become part of the verb itself. "I might *not*."

Participles are words formed from verbs that are often used to modify a noun, noun phrase, verb, or verb phrase.

The *running* teenager collided with the cyclist.

Participles can also create compound verb forms.

He is *speaking*.

Verbs have five basic forms: the **base** form, the **-s** form, the **-ing** form, the **past** form, and the **past participle** form.

The past forms are either **regular** (*love/loved; hate/hated*) or **irregular** because they don't end by adding the common past tense suffix "-ed" (*go/went; fall/fell; set/set*).

Verb Forms

Shifting verb forms entails **conjugation,** which is used to indicate tense, voice, or mood.

Verb tense is used to show when the action in the sentence took place. There are several different verb tenses, and it is important to know how and when to use them. Some verb tenses can be achieved by changing the form of the verb, while others require the use of helping verbs (e.g., *is, was,* or *has*).

Present tense shows the action is happening currently or is ongoing:

I walk to work every morning.

She is stressed about the deadline.

Past tense shows that the action happened in the past or that the state of being is in the past:

I walked to work yesterday morning.

She was stressed about the deadline.

Future tense shows that the action will happen in the future or is a future state of being:

I will walk to work tomorrow morning.

She will be stressed about the deadline.

Present perfect tense shows action that began in the past, but continues into the present:

I have walked to work all week.

She has been stressed about the deadline.

Past perfect tense shows an action was finished before another took place:

I had walked all week until I sprained my ankle.

She had been stressed about the deadline until we talked about it.

Future perfect tense shows an action that will be completed at some point in the future:

By the time the bus arrives, I will have walked to work already.

Voice

Verbs can be in the active or passive voice. When the subject completes the action, the verb is in **active voice**. When the subject receives the action of the sentence, the verb is in **passive voice**.

Active: Jamie ate the ice cream.

Passive: The ice cream was eaten by Jamie.

In active voice, the subject (*Jamie*) is the "do-er" of the action (*ate*). In passive voice, the subject *ice cream* receives the action of being eaten.

While passive voice can add variety to writing, active voice is the generally preferred sentence structure.

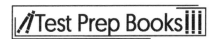

Mood

Mood is used to show the speaker's feelings about the subject matter. In English, there is indicative mood, imperative mood, and subjective mood.

Indicative mood is used to state facts, ask questions, or state opinions:

Bob will make the trip next week.

When can Bob make the trip?

Imperative mood is used to state a command or make a request:

Wait in the lobby.

Please call me next week.

Subjunctive mood is used to express a wish, an opinion, or a hope that is contrary to fact:

If I were in charge, none of this would have happened.

Allison wished she could take the exam over again when she saw her score.

Adjectives

Adjectives are words used to modify nouns and pronouns. They can be used alone or in a series and are used to further define or describe the nouns they modify.

Mark made us a delicious, four-course meal.

The words *delicious* and *four-course* are adjectives that describe the kind of meal Mark made.

Articles are also considered adjectives because they help to describe nouns. Articles can be general or specific. The three articles in English are: *a, an,* and *the.*

Indefinite articles *(a, an)* are used to refer to nonspecific nouns. The article *a* proceeds words beginning with consonant sounds, and the article *an* proceeds words beginning with vowel sounds.

A car drove by our house.

An alligator was loose at the zoo.

He has always wanted a ukulele. (The first *u* makes a *y* sound.)

Note that *a* and *an* should only proceed nonspecific nouns that are also singular. If a nonspecific noun is plural, it does not need a proceeding article.

Alligators were loose at the zoo.

The **definite article** *(the)* is used to refer to specific nouns:

The car pulled into our driveway.

Note that *the* should proceed all specific nouns regardless of whether they are singular or plural.

The cars pulled into our driveway.

Comparative adjectives are used to compare nouns. When they are used in this way, they take on positive, comparative, or superlative form.

The **positive** form is the normal form of the adjective:

Alicia is tall.

The **comparative** form shows a comparison between two things:

Alicia is taller than Maria.

Superlative form shows comparison between more than two things:

Alicia is the tallest girl in her class.

Usually, the comparative and superlative can be made by adding –*er* and –*est* to the positive form, but some verbs call for the helping verbs *more* or *most*. Other exceptions to the rule include adjectives like *bad*, which uses the comparative *worse* and the superlative *worst*.

An **adjective phrase** is not a bunch of adjectives strung together, but a group of words that describes a noun or pronoun and, thus, functions as an adjective. *Very happy* is an adjective phrase; so are *way too hungry* and *passionate about traveling*.

Adverbs

Adverbs have more functions than adjectives because they modify or qualify verbs, adjectives, or other adverbs as well as word groups that express a relation of place, time, circumstance, or cause. Therefore, adverbs answer any of the following questions: *How, when, where, why, in what way, how often, how much, in what condition*, and/or *to what degree. How good looking is he? He is <u>very</u> handsome.*

Here are some examples of adverbs for different situations:

- how: quickly
- when: daily
- where: there
- in what way: easily
- how often: often
- how much: much
- in what condition: badly
- what degree: hardly

As one can see, for some reason, many adverbs end in -*ly*.

Adverbs do things like emphasize (*really, simply,* and *so*), amplify (*heartily, completely,* and *positively*), and tone down (*almost, somewhat,* and *mildly*).

Adverbs also come in phrases.

The dog ran as <u>though his life depended on it.</u>

Prepositions

Prepositions are connecting words and, while there are only about 150 of them, they are used more often than any other individual groups of words. They describe relationships between other words. They are placed before a noun or pronoun, forming a phrase that modifies another word in the sentence. **Prepositional phrases** begin with a preposition and end with a noun or pronoun, the **object of the preposition.** *A pristine lake is <u>near the store</u> and <u>behind the bank.</u>*

Some commonly used prepositions are *about, after, anti, around, as, at, behind, beside, by, for, from, in, into, of, off, on, to,* and *with.*

Complex prepositions, which also come before a noun or pronoun, consist of two or three words such as *according to, in regards to,* and *because of.*

Conjunctions

Conjunctions are vital words that connect words, phrases, thoughts, and ideas. Conjunctions show relationships between components. There are two types:

Coordinating conjunctions are the primary class of conjunctions placed between words, phrases, clauses, and sentences that are of equal grammatical rank; the coordinating conjunctions are *for, and, nor, but, or, yet,* and *so.* A useful memorization trick is to remember that the first letter of these conjunctions collectively spell the word *fanboys.*

> I need to go shopping, *but* I must be careful to leave enough money in the bank.
> She wore a black, red, *and* white shirt.

Subordinating conjunctions are the secondary class of conjunctions. They connect two unequal parts, one **main** (or **independent**) and the other *subordinate* (or *dependent*). I must go to the store *even though* I do not have enough money in the bank.

> *Because* I read the review, I do not want to go to the movie.

Notice that the presence of subordinating conjunctions makes clauses dependent. *I read the review* is an independent clause, but *because* makes the clause dependent. Thus, it needs an independent clause to complete the sentence.

Interjections

Interjections are words used to express emotion. Examples include *wow, ouch,* and *hooray.* Interjections are often separate from sentences; in those cases, the interjection is directly followed by an exclamation point. In other cases, the interjection is included in a sentence and followed by a comma. The punctuation plays a big role in the intensity of the emotion that the interjection is expressing. Using a comma or semicolon indicates less excitement than using an exclamation mark.

Capitalization Rules

Here's a non-exhaustive list of things that should be capitalized.

- The first word of every sentence

- The first word of every line of poetry

- The first letter of proper nouns (World War II)

- Holidays (Valentine's Day)

- The days of the week and months of the year (Tuesday, March)

- The first word, last word, and all major words in the titles of books, movies, songs, and other creative works (In the novel, To Kill a Mockingbird, note that a is lowercase since it's not a major word, but to is capitalized since it's the first word of the title)

- Titles when preceding a proper noun (President Roberto Gonzales, Aunt Judy)

When simply using a word such as president or secretary, though, the word is not capitalized.

Officers of the new business must include a *president* and *treasurer*.

Seasons—spring, fall, etc.—are not capitalized.

North, *south*, *east*, and *west* are capitalized when referring to regions but are not when being used for directions. In general, if it's preceded by *the* it should be capitalized.

I'm from the South.
I drove south.

End Punctuation

Periods (.) are used to end a sentence that is a statement (**declarative**) or a command (**imperative**). They should not be used in a sentence that asks a question or is an exclamation. Periods are also used in abbreviations, which are shortened versions of words.

- Declarative: The boys refused to go to sleep.
- Imperative: Walk down to the bus stop.
- Abbreviations: Joan Roberts, M.D., Apple Inc., Mrs. Adamson
- If a sentence ends with an abbreviation, it is inappropriate to use two periods. It should end with a single period after the abbreviation.

The chef gathered the ingredients for the pie, which included apples, flour, sugar, etc.

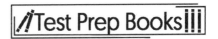

Question marks *(?)* are used with direct questions (**interrogative**). An **indirect question** can use a period:

> Interrogative: When does the next bus arrive?

> Indirect Question: I wonder when the next bus arrives.

An **exclamation point** *(!)* is used to show strong emotion or can be used as an interjection. This punctuation should be used sparingly in formal writing situations.

> What an amazing shot!

> Whoa!

Commas

A **comma** (,) is the punctuation mark that signifies a pause—breath—between parts of a sentence. It denotes a break of flow. Proper comma usage helps readers understand the writer's intended emphasis of ideas.

In a complex sentence—one that contains a **subordinate** (dependent) clause or clauses—the use of a comma is dictated by where the subordinate clause is located. If the subordinate clause is located before the main clause, a comma is needed between the two clauses.

> I will not pay for the steak, *because I don't have that much money*.

Generally, if the subordinate clause is placed after the main clause, no punctuation is needed.

> I did well on my exam because I studied two hours the night before.

Notice how the last clause is dependent because it requires the earlier independent clauses to make sense.

Use a comma on both sides of an interrupting phrase.

> I will pay for the ice cream, *chocolate and vanilla*, and then will eat it all myself.

The words forming the phrase in italics are nonessential (extra) information. To determine if a phrase is nonessential, try reading the sentence without the phrase and see if it's still coherent.

A comma is not necessary in this next sentence because no interruption—nonessential or extra information—has occurred. Read sentences aloud when uncertain.

> I will pay for his chocolate and vanilla ice cream and then will eat it all myself.

If the nonessential phrase comes at the beginning of a sentence, a comma should only go at the end of the phrase. If the phrase comes at the end of a sentence, a comma should only go at the beginning of the phrase.

Other types of interruptions include the following:

- interjections: Oh no, I am not going.
- abbreviations: Barry Potter, M.D., specializes in heart disorders.
- direct addresses: Yes, Claudia, I am tired and going to bed.
- parenthetical phrases: His wife, lovely as she was, was not helpful.
- transitional phrases: Also, it is not possible.

The second comma in the following sentence is called an **Oxford comma**.

I will pay for ice cream, syrup, and pop.

It is a comma used after the second-to-last item in a series of three or more items. It comes before the word *or* or *and*. Not everyone uses the Oxford comma; it is optional, but many believe it is needed. The comma functions as a tool to reduce confusion in writing. So, if omitting the Oxford comma would cause confusion, then it's best to include it.

Commas are used in math to mark the place of thousands in numerals, breaking them up so they are easier to read. Other uses for commas are in dates (*March 19, 2016*), letter greetings (*Dear Sally,*), and in between cities and states (*Louisville, KY*).

Semicolons

A **semicolon** *(;)* is used to connect ideas in a sentence in some way. There are three main ways to use semicolons.

Link two independent clauses without the use of a coordinating conjunction:

I was late for work again; I'm definitely going to get fired.

Link two independent clauses with a transitional word:

The songs were all easy to play; therefore, he didn't need to spend too much time practicing.

Between items in a series that are already separated by commas or if necessary to separate lengthy items in a list:

Starbucks has locations in Media, PA; Swarthmore, PA; and Morton, PA.

Several classroom management issues presented in the study: the advent of a poor teacher persona in the context of voice, dress, and style; teacher follow-through from the beginning of the school year to the end; and the depth of administrative support, including ISS and OSS protocol.

Colons

A **colon** (:) is used after an independent clause to present an explanation or draw attention to what comes next in the sentence. There are several uses.

Explanations of ideas:

They soon learned the hardest part about having a new baby: sleep deprivation.

Lists of items:

> Shari picked up all the supplies she would need for the party: cups, plates, napkins, balloons, streamers, and party favors.

Time, subtitles, general salutations:

> The time is 7:15.

> I read a book entitled *Pluto: A Planet No More*.

> To whom it may concern:

Parentheses and Dashes

Parentheses are half-round brackets that look like this: ( ). They set off a word, phrase, or sentence that is an afterthought, explanation, or side note relevant to the surrounding text but not essential. A pair of commas is often used to set off this sort of information, but parentheses are generally used for information that would not fit well within a sentence or that the writer deems not important enough to be structurally part of the sentence.

> The picture of the heart (see above) shows the major parts you should memorize.
> Mount Everest is one of three mountains in the world that are over 28,000 feet high (K2 and Kanchenjunga are the other two).

See how the sentences above are complete without the parenthetical statements? In the first example, *see above* would not have fit well within the flow of the sentence. The second parenthetical statement could have been a separate sentence, but the writer deemed the information not pertinent to the topic.

The **em-dash** (—) is a mark longer than a hyphen used as a punctuation mark in sentences and to set apart a relevant thought. Even after plucking out the line separated by the dash marks, the sentence will be intact and make sense.

> Looking out the airplane window at the landmarks—Lake Clarke, Thompson Community College, and the bridge—she couldn't help but feel excited to be home.

The dashes use is similar to that of parentheses or a pair of commas. So, what's the difference? Many believe that using dashes makes the clause within them stand out while using parentheses is subtler. It's advised to not use dashes when commas could be used instead.

Ellipses

An **ellipsis** (...) consists of three handy little dots that can speak volumes on behalf of irrelevant material. Writers use them in place of words, lines, phrases, list content, or paragraphs that might just as easily have been omitted from a passage of writing. This can be done to save space or to focus only on the specifically relevant material.

> Exercise is good for some unexpected reasons. Watkins writes, "Exercise has many benefits such as...reducing cancer risk."

In the example above, the ellipsis takes the place of the other benefits of exercise that are more expected.

The ellipsis may also be used to show a pause in sentence flow.

> "I'm wondering...how this could happen," Dylan said in a soft voice.

Quotation Marks

Double quotation marks are used at the beginning and end of a direct quote. They are also used with certain titles and to indicate that a term being used is slang or referenced in the sentence. Quotation marks should not be used with an indirect quote. Single quotation marks are used to indicate a quote within a quote.

> Direct quote: "The weather is supposed to be beautiful this week," she said.

> Indirect quote: One of the customers asked if the sale prices were still in effect.

> Quote within a quote: "My little boy just said 'Mama, I want cookie,'" Maria shared.

Titles: Quotation marks should also be used to indicate titles of short works or sections of larger works, such as chapter titles. Other works that use quotation marks include poems, short stories, newspaper articles, magazine articles, web page titles, and songs.

> "The Road Not Taken" is my favorite poem by Robert Frost.

> "What a Wonderful World" is one of my favorite songs.

Specific or emphasized terms: Quotation marks can also be used to indicate a technical term or to set off a word that is being discussed in a sentence. Quotation marks can also indicate sarcasm.

> The new step, called "levigation," is a very difficult technique.

> He said he was "hungry" multiple times, but he only ate two bites.

Use with other punctuation: The use of quotation marks with other punctuation varies, depending on the role of the ending or separating punctuation.

In American English, periods and commas always go inside the quotation marks:

> "This is the last time you are allowed to leave early," his boss stated.

> The newscaster said, "We have some breaking news to report."

Question marks or exclamation points go inside the quotation marks when they are part of a direct quote:

> The doctor shouted, "Get the crash cart!"

When the question mark or exclamation point is part of the sentence, not the quote, it should be placed outside of the quotation marks:

> Was it Jackie that said, "Get some potatoes at the store"?

Apostrophes

This punctuation mark, the **apostrophe** (') is a versatile mark. It has several different functions:

- Quotes: Apostrophes are used when a second quote is needed within a quote.

    o In my letter to my friend, I wrote, "The girl had to get a new purse, and guess what Mary did? She said, 'I'd like to go with you to the store.' I knew Mary would buy it for her."

- Contractions: Another use for an apostrophe in the quote above is a contraction. *I'd is used for I would.*

- Possession: An apostrophe followed by the letter s shows possession (Mary's purse). If the possessive word is plural, the apostrophe generally just follows the word. Not all possessive pronouns require apostrophes.

    o The trees' leaves are all over the ground.

Hyphens

The **hyphen** (-) is a small hash mark that can be used to join words to show that they are linked.

Hyphenate two words that work together as a single adjective (a compound adjective).

honey-covered biscuits

Some words always require hyphens, even if not serving as an adjective.

merry-go-round

Hyphens always go after certain prefixes like *anti-* & *all-*.

Hyphens should also be used when the absence of the hyphen would cause a strange vowel combination (*semi-engineer*) or confusion. For example, *re-collect* should be used to describe something being gathered twice rather than being written as *recollect*, which means to remember.

Subjects

Every sentence must include a subject and a verb. The **subject** of a sentence is who or what the sentence is about. It's often directly stated and can be determined by asking "Who?" or "What?" did the action:

Most sentences contain a **direct subject**, in which the subject is mentioned in the sentence.

Kelly mowed the lawn.

Who mowed the lawn? *Kelly*

The air-conditioner ran all night

What ran all night? *the air-conditioner*

The subject of imperative sentences is the implied *you*, because imperative subjects are commands:

> Go home after the meeting.

> Who should go home after the meeting? *you* (implied)

In **expletive sentences** that start with "there are" or "there is," the subject is found after the predicate. The subject cannot be "there," so it must be another word in the sentence:

> There is a cup sitting on the coffee table.

> What is sitting on the coffee table? *a cup*

Simple and Complete Subjects

A **complete subject** includes the simple subject and all the words modifying it, including articles and adjectives. A **simple subject** is the single noun without its modifiers.

> A warm, chocolate-chip cookie sat on the kitchen table.

> Complete subject: *a warm, chocolate-chip cookie*

> Simple subject: *cookie*

The words *a, warm, chocolate,* and *chip* all modify the simple subject *cookie.*

There might also be a **compound subject**, which would be two or more nouns without the modifiers.

> A little girl and her mother walked into the shop.

> Complete subject: *A little girl and her mother*

> Compound subject: *girl, mother*

In this case, *the girl and her mother* are both completing the action of walking into the shop, so this is a compound subject.

Predicates

In addition to the subject, a sentence must also have a predicate. The **predicate** contains a verb and tells something about the subject. In addition to the verb, a predicate can also contain a direct or indirect object, object of a preposition, and other phrases.

> The cats napped on the front porch.

In this sentence, cats is the subject because the sentence is about cats.

The **complete predicate** is everything else in the sentence: *napped on the front porch.* This phrase is the predicate because it tells us what the cats did.

This sentence can be broken down into a simple subject and predicate:

> Cats napped.

In this sentence, *cats* is the simple subject, and *napped* is the **simple predicate**.

Although the sentence is very short and doesn't offer much information, it's still considered a complete sentence because it contains a subject and predicate.

Like a compound subject, a sentence can also have a **compound predicate**. This is when the subject is or does two or more things in the sentence.

> This easy chair reclines and swivels.

In this sentence, *this easy chair* is the complete subject. *Reclines and swivels* shows two actions of the chair, so this is the compound predicate.

Subject-Verb Agreement

The subject of a sentence and its verb must agree. The cornerstone rule of subject-verb agreement is that subject and verb must agree in number. Whether the subject is singular or plural, the verb must follow suit.

> Incorrect: The houses is new.
> Correct: The houses are new.
> Also Correct: The house is new.

In other words, a singular subject requires a singular verb; a plural subject requires a plural verb. The words or phrases that come between the subject and verb do not alter this rule.

> Incorrect: The houses built of brick is new.
> Correct: The houses built of brick are new.

> Incorrect: The houses with the sturdy porches is new.
> Correct: The houses with the sturdy porches are new.

The subject will always follow the verb when a sentence begins with *here* or *there.* Identify these with care.

> Incorrect: Here *is* the *houses* with sturdy porches.
> Correct: Here *are* the *houses* with sturdy porches.

The subject in the sentences above is not *here*, it is *houses*. Remember, *here* and *there* are never subjects. Be careful that contractions such as *here's* or *there're* do not cause confusion!

Two subjects joined by *and* require a plural verb form, except when the two combine to make one thing:

> Incorrect: Garrett and Jonathan is over there.
> Correct: Garrett and Jonathan are over there.

> Incorrect: Spaghetti and meatballs are a delicious meal!
> Correct: Spaghetti and meatballs is a delicious meal!

In the example above, *spaghetti and meatballs* is a compound noun. However, *Garrett and Jonathan* is not a compound noun.

Two singular subjects joined by *or, either/or,* or *neither/nor* call for a singular verb form.

> Incorrect: Butter or syrup are acceptable.
> Correct: Butter or syrup is acceptable.

Plural subjects joined by *or, either/or,* or *neither/nor* are, indeed, plural.

> The chairs or the boxes are being moved next.

If one subject is singular and the other is plural, the verb should agree with the closest noun.

> Correct: The chair or the boxes are being moved next.
> Correct: The chairs or the box is being moved next.

Some plurals of money, distance, and time call for a singular verb.

> Incorrect: Three dollars *are* enough to buy that.
> Correct: Three dollars *is* enough to buy that.

For words declaring degrees of quantity such as *many of, some of,* or *most of,* let the noun that follows of be the guide:

> Incorrect: Many of the books is in the shelf.
> Correct: Many of the books are in the shelf.

> Incorrect: Most of the pie *are* on the table.
> Correct: Most of the pie *is* on the table.

For indefinite pronouns like *anybody* or *everybody*, use singular verbs.

> Everybody *is* going to the store.

However, the pronouns *few, many, several, all, some,* and *both* have their own rules and use plural forms.

> Some *are* ready.

Some nouns like *crowd* and *congress* are called **collective nouns** and they require a singular verb form.

> Congress *is* in session.
> The news *is* over.

Books and movie titles, though, including plural nouns such as *Great Expectations*, also require a singular verb. Remember that only the subject affects the verb. While writing tricky subject-verb arrangements, say them aloud. Listen to them. Once the rules have been learned, one's ear will become sensitive to them, making it easier to pick out what's right and what's wrong.

Direct Objects

The **direct object** is the part of the sentence that receives the action of the verb. It is a noun and can usually be found after the verb. To find the direct object, first find the verb, and then ask the question *who* or *what* after it.

>The bear climbed the tree.

>What did the bear climb? *the tree*

Indirect Objects

An **indirect object** receives the direct object. It is usually found between the verb and the direct object. A strategy for identifying the indirect object is to find the verb and ask the questions *to whom/for whom* or *to what/for what*.

>Jane made her daughter a cake.

>For whom did Jane make the cake? *her daughter*

Cake is the direct object because it is what Jane made, and *daughter* is the indirect object because she receives the cake.

Complements

A **complement** completes the meaning of an expression. A complement can be a pronoun, noun, or adjective. A **verb complement** refers to the direct object or indirect object in the sentence. An **object complement** gives more information about the direct object:

>The magician got the kids excited.

Kids is the direct object, and *excited* is the object complement.

A **subject complement** comes after a linking verb. It is typically an adjective or noun that gives more information about the subject:

>The king was noble and spared the thief's life.

Noble describes the *king* and follows the linking verb *was*.

Predicate Nouns

A **predicate noun** renames the subject:

>John is a carpenter.

The subject is *John*, and the predicate noun is *carpenter*.

.ate.

Predicate Adjectives

A **predicate adjective** describes the subject:

Margaret is beautiful.

The subject is *Margaret*, and the predicate adjective is *beautiful*.

Homonyms

Homonyms are words that sound the same but are spelled differently, and they have different meanings. There are several common homonyms that give writers trouble.

There, They're, and *Their*
The word *there* can be used as an adverb, adjective, or pronoun:

There are ten children on the swim team this summer.

I put my book over *there*, but now I can't find it.

The word *they're* is a contraction of the words *they* and *are*:

They're flying in from Texas on Tuesday.

The word *their* is a possessive pronoun:

I store *their* winter clothes in the attic.

Its and *It's*
Its is a possessive pronoun:

The cat licked *its* injured paw.

It's is the contraction for the words *it* and *is*:

It's unbelievable how many people opted not to vote in the last election.

Your and You're
Your is a possessive pronoun:

Can I borrow *your* lawnmower this weekend?

You're is a contraction for the words *you* and *are*:

You're about to embark on a fantastic journey.

To, Too, and *Two*
To is an adverb or a preposition used to show direction, relationship, or purpose:

We are going *to* New York.

They are going *to* see a show.

Too is an adverb that means more than enough, also, and very:

> You have had *too* much candy.

> We are on vacation that week, *too*.

Two is the written-out form of the numeral 2:

> *Two* of the shirts didn't fit, so I will have to return them.

New and Knew
New is an adjective that means recent:

> There's a *new* customer on the phone.

Knew is the past tense of the verb *know*:

> I *knew* you'd have fun on this ride.

Affect and Effect
Affect and *effect* are complicated because they are used as both nouns and verbs, have similar meanings, and are pronounced the same.

	Affect	**Effect**
Noun Definition	emotional state	result
Noun Example	The patient's affect was flat.	The effects of smoking are well documented.
Verb Definition	to influence	to bring about
Verb Example	The pollen count affects my allergies.	The new candidate hopes to effect change.

Independent and Dependent Clauses

Independent and **dependent** clauses are strings of words that contain both a subject and a verb. An independent clause *can* stand alone as complete thought, but a dependent clause *cannot*. A dependent clause relies on other words to be a complete sentence.

> Independent clause: The keys are on the counter.
> Dependent clause: If the keys are on the counter

Notice that both clauses have a subject (*keys*) and a verb (*are*). The independent clause expresses a complete thought, but the word *if* at the beginning of the dependent clause makes it *dependent* on other words to be a complete thought.

> Independent clause: If the keys are on the counter, please give them to me.

This example constitutes a complete sentence since it includes at least one verb and one subject and is a complete thought. In this case, the independent clause has two subjects (*keys* & an implied *you*) and two verbs (*are* & *give*).

> Independent clause: I went to the store.
> Dependent clause: Because we are out of milk,
>
> Complete Sentence: Because we are out of milk, I went to the store.
> Complete Sentence: I went to the store because we are out of milk.

Phrases

A **phrase** is a group of words that do not make a complete thought or a clause. They are parts of sentences or clauses. Phrases can be used as nouns, adjectives, or adverbs. A phrase does not contain both a subject and a verb.

Prepositional Phrases

A **prepositional phrase** shows the relationship between a word in the sentence and the object of the preposition. The **object of the preposition** is a noun that follows the preposition.

> The orange pillows are on the couch.

On is the preposition, and *couch* is the object of the preposition.

> She brought her friend with the nice car.

With is the preposition, and *car* is the object of the preposition. Here are some common prepositions:

about	as	at	after
by	for	from	in
of	on	to	with

Verbals and Verbal Phrases

Verbals are forms of verbs that act as other parts of speech. They can be used as nouns, adjectives, or adverbs. Though they are verb forms, they are not to be used as the verb in the sentence. A word group that is based on a verbal is considered a **verbal phrase**. There are three major types of verbals: participles, gerunds, and infinitives.

Participles are verbals that act as adjectives. The present participle ends in *–ing*, and the past participle ends in *–d, -ed, -n, or-t*.

Verb	Present Participle	Past Participle
walk	walking	walked
share	sharing	shared

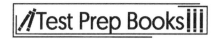

Participial phrases are made up of the participle and modifiers, complements, or objects.

> Crying for most of an hour, the baby didn't seem to want to nap.

> Having already taken this course, the student was bored during class.

> *Crying for most of an hour* and *Having already taken this course* are the participial phrases.

Gerunds are verbals that are used as nouns and end in *–ing*. A gerund can be the subject or object of the sentence like a noun. Note that a present participle can also end in *–ing*, so it is important to distinguish between the two. The gerund is used as a noun, while the participle is used as an adjective.

> Swimming is my favorite sport.

> I wish I were sleeping.

A **gerund phrase** includes the gerund and any modifiers or complements, direct objects, indirect objects, or pronouns.

> Cleaning the house is my least favorite weekend activity

Cleaning the house is the gerund phrase acting as the subject of the sentence.

> The most important goal this year is raising money for charity.

Raising money for charity is the gerund phrase acting as the direct object.

> The police accused the woman of stealing the car.

The gerund phrase *stealing the car* is the object of the preposition in this sentence.

An **infinitive** is a verbal made up of the word *to* and a verb. Infinitives can be used as nouns, adjectives, or adverbs.

> Examples: To eat, to jump, to swim, to lie, to call, to work

An *infinitive phrase* is made up of the infinitive plus any complements or modifiers. The infinitive phrase *to wait* is used as the subject in this sentence:

> To wait was not what I had in mind.

The infinitive phrase *to sing* is used as the subject complement in this sentence:

> Her dream is to sing.

The infinitive phrase *to grow* is used as an adverb in this sentence:

> Children must eat to grow.

Appositive Phrases

An **appositive** is a noun or noun phrase that renames a noun that comes immediately before it in the sentence. An appositive can be a single word or several words. These phrases can be **essential** or **nonessential**. An essential appositive phrase is necessary to the meaning of the sentence and a nonessential appositive phrase is not. It is important to be able to distinguish these for purposes of comma use.

Essential: My sister Christina works at a school.

Naming which sister is essential to the meaning of the sentence, so no commas are needed.

Nonessential: My sister, who is a teacher, is coming over for dinner tonight.

Who is a teacher is not essential to the meaning of the sentence, so commas are required.

Absolute Phrases

An **absolute phrase** modifies a noun without using a conjunction. It is not the subject of the sentence and is not a complete thought on its own. Absolute phrases are set off from the independent clause with a comma.

Arms outstretched, she yelled at the sky.

All things considered, this has been a great day.

The Four Types of Sentence Structures

A **simple sentence** has one independent clause.

I am going to win.

A **compound sentence** has two independent clauses. A conjunction—*for, and, nor, but, or, yet, so*—links them together. Note that each of the independent clauses has a subject and a verb.

I am going to win, but the odds are against me.

A **complex sentence** has one independent clause and one or more dependent clauses.

I am going to win, even though I don't deserve it.

Even though I don't deserve it is a dependent clause. It does not stand on its own. Some conjunctions that link an independent and a dependent clause are *although, because, before, after, that, when, which*, and *while*.

A **compound-complex sentence** has at least three clauses, two of which are independent and at least one that is a dependent clause.

While trying to dance, I tripped over my partner's feet, but I regained my balance quickly.

The dependent clause is *While trying to dance.*

Sentence Fragments

A **sentence fragment** is an incomplete sentence. An independent clause is made up of a subject and a predicate, and both are needed to make a complete sentence.

Sentence fragments often begin with relative pronouns (when, which), subordinating conjunctions (because, although) or gerunds (trying, being, seeing). They might be missing the subject or the predicate.

The most common type of fragment is the isolated dependent clause, which can be corrected by joining it to the independent clause that appears before or after the fragment:

Fragment: While the cookies baked.

Correction: While the cookies baked, we played cards. (We played cards while the cookies baked.)

Run-on Sentences

A **run-on sentence** is created when two independent clauses (complete thoughts) are joined without correct punctuation or a conjunction. Run-on sentences can be corrected in the following ways:

- Join the independent clauses with a comma and coordinating conjunction.

  - Run-on: We forgot to return the library books we had to pay a fine.

  - Correction: We forgot to return the library books, so we had to pay a fine.

- Join the independent clauses with a semicolon, dash, or colon when the clauses are closely related in meaning.

  - Run-on: I had a salad for lunch every day this week I feel healthier already.

  - Correction: I had a salad for lunch every day this week; I feel healthier already.

- Join the independent clauses with a semicolon and a conjunctive adverb.

  - Run-on: We arrived at the animal shelter on time however the dog had already been adopted.

  - Correction: We arrived at the animal shelter on time; however, the dog had already been adopted.

- Separate the independent clauses into two sentences with a period.

  - Run-on: He tapes his favorite television show he never misses an episode.

  - Correction: He tapes his favorite television show. He never misses an episode.

- Rearrange the wording of the sentence to create an independent clause and a dependent clause.

  - Run-on: My wedding date is coming up I am getting more excited to walk down the aisle.

  - Correction: As my wedding date approaches, I am getting more excited to walk down the aisle.

Dangling and Misplaced Modifiers

A **modifier** is a phrase that describes, alters, limits, or gives more information about a word in the sentence. The two most common issues are dangling and misplaced modifiers.

A **dangling modifier** is created when the phrase modifies a word that is not clearly stated in the sentence.

Dangling modifier: Having finished dinner, the dishes were cleared from the table.

Correction: Having finished dinner, Amy cleared the dishes from the table.

In the first sentence, *having finished dinner* appears to modify *the dishes*, which obviously can't finish dinner. The second sentence adds the subject *Amy*, to make it clear who has finished dinner.

Dangling modifier: Hoping to improve test scores, all new books were ordered for the school.

Correction: Hoping to improve test scores, administrators ordered all new books for the school.

Without the subject *administrators*, it appears the books are hoping to improve test scores, which doesn't make sense.

Misplaced modifiers are placed incorrectly in the sentence, which can cause confusion. Compare these examples:

Misplaced modifier: Rory purchased a new flat screen television and placed it on the wall above the fireplace, with all the bells and whistles.

Revised: Rory purchased a new flat screen television, with all the bells and whistles, and placed it on the wall above the fireplace.

The bells and whistles should modify the television, not the fireplace.

Misplaced modifier: The delivery driver arrived late with the pizza, who was usually on time.

Revised: The delivery driver, who usually was on time, arrived late with the pizza.

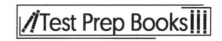

This suggests that the delivery driver was usually on time, instead of the pizza.

> Misplaced modifier: We saw a family of ducks on the way to church.

> Revised: On the way to church, we saw a family of ducks.

The misplaced modifier, here, suggests the *ducks* were on their way to church, instead of the pronoun *we*.

Split Infinitives

An **infinitive** is made up of the word *to* and a verb, such as: to run, to jump, to ask. A **split infinitive** is created when a word comes between *to* and the verb.

> Split infinitive: To quickly run

> Correction: To run quickly

> Split infinitive: To quietly ask

> Correction: To ask quietly

Double Negatives

A **double negative** is a negative statement that includes two negative elements. This is incorrect in Standard English.

> Incorrect: She hasn't never come to my house to visit.

> Correct: She has never come to my house to visit.

The intended meaning is that she has never come to the house, so the double negative is incorrect. However, it is possible to use two negatives to create a positive statement.

> Correct: She was not unhappy with her performance on the quiz.

In this case, the double negative, *was not unhappy*, is intended to show a positive, so it is correct. This means that she was somewhat happy with her performance.

Faulty Parallelism

It is necessary to use parallel construction in sentences that have multiple similar ideas. Using parallel structure provides clarity in writing. **Faulty parallelism** is created when multiple ideas are joined using different sentence structures. Compare these examples:

> Incorrect: We start each practice with stretches, a run, and fielding grounders.
> Correct: We start each practice with stretching, running, and fielding grounders.

> Incorrect: I watched some television, reading my book, and fell asleep.
> Correct: I watched some television, read my book, and fell asleep.

Incorrect: Some of the readiness skills for Kindergarten are to cut with scissors, to tie shoes, and dressing independently.

Correct: Some of the readiness skills for Kindergarten are being able to cut with scissors, to tie shoes, and to dress independently.

Subordination

If multiple pieces of information in a sentence are not equal, they can be joined by creating an independent clause and a dependent clause. The less important information becomes the **subordinate clause:**

Draft: The hotel was acceptable. We wouldn't stay at the hotel again.

Revised: Though the hotel was acceptable, we wouldn't stay there again.

The more important information (*we wouldn't stay there again*) becomes the main clause, and the less important information (*the hotel was acceptable*) becomes the subordinate clause.

Practice Questions

Questions 1–9 are based on the following passage:

While all dogs (1) descend through gray wolves, it's easy to notice that dog breeds come in a variety of shapes and sizes. With such a (2) drastic range of traits, appearances and body types dogs are one of the most variable and adaptable species on the planet. But why so many differences. The answer is that humans have actually played a major role in altering the biology of dogs. This was done through a process called selective breeding.

(3) Selective breeding which is also called artificial selection is the process in which animals with desired traits are bred in order to produce offspring that share the same traits. In natural evolution, (4) animals must adapt to their environments increase their chance of survival. Over time, certain traits develop in animals that enable them to thrive in these environments. Those animals with more of these traits, or better versions of these traits, gain an (5) advantage over others of their species. Therefore, the animal's chances to mate are increased and these useful genes are passed onto their offspring. With dog breeding, humans select traits that are desired and encourage more of these desired traits in other dogs by breeding dogs that already have them.

The reason for different breeds of dogs is that there were specific needs that humans wanted to fill with their animals. For example, scent hounds are known for their extraordinary ability to track game through scent. These breeds are also known for their endurance in seeking deer and other prey. Therefore, early hunters took dogs that displayed these abilities and bred them to encourage these traits. Later generations took on characteristics that aided these desired traits.

1. Which of the following would be the best choice for this sentence (reproduced below)?

While all dogs (1) descend through gray wolves, it's easy to notice that dog breeds come in a variety of shapes and sizes.

    a. NO CHANGE
    b. descend by gray wolves
    c. descend from gray wolves
    d. descended through gray wolves

2. Which of the following would be the best choice for this sentence (reproduced below)?

With such a (2) drastic range of traits, appearances and body types, dogs are one of the most variable and adaptable species on the planet.

    a. NO CHANGE
    b. drastic range of traits, appearances, and body types,
    c. drastic range of traits and appearances and body types,
    d. drastic range of traits, appearances, as well as body types,

3. Which of the following would be the best choice for this sentence (reproduced below)?

(3) <u>Selective breeding which is also called artificial selection is the process</u> in which animals with desired traits are bred in order to produce offspring that share the same traits.

a. NO CHANGE
b. Selective breeding, which is also called artificial selection is the process
c. Selective breeding which is also called, artificial selection, is the process
d. Selective breeding, which is also called artificial selection, is the process

4. Which of the following would be the best choice for this sentence (reproduced below)?

In natural evolution, (4) <u>animals must adapt to their environments increase their chance of survival.</u>

a. NO CHANGE
b. animals must adapt to their environments to increase their chance of survival.
c. animals must adapt to their environments, increase their chance of survival.
d. animals must adapt to their environments, increasing their chance of survival.

5. Which of the following would be the best choice for this sentence (reproduced below)?

Those animals with more of these traits, or better versions of these traits, gain an (5) <u>advantage over others of their species.</u>

a. NO CHANGE
b. advantage over others, of their species.
c. advantages over others of their species.
d. advantage over others.

Answer Explanations

1. C: Choice *C* correctly uses *from* to describe the fact that dogs are related to wolves. The word *through* is incorrectly used here, so Choice *A* is incorrect. Choice *B* makes no sense. Choice *D* unnecessarily changes the verb tense in addition to incorrectly using *through*.

2. B: Choice *B* is correct because the Oxford comma is applied, clearly separating the specific terms. Choice *A* lacks this clarity. Choice *C* is correct but too wordy since commas can be easily applied. Choice *D* doesn't flow with the sentence's structure.

3. D: Choice *D* is correct because the commas serve to distinguish that *artificial selection* is just another term for *selective breeding* before the sentence continues. The structure is preserved, and the sentence can flow with more clarity. Choice *A* is incorrect because the sentence needs commas to avoid being a run-on. Choice *B* is close but still lacks the required comma after *selection*, so this is incorrect. Choice *C* is incorrect because the comma to set off the aside should be placed after *breeding* instead of *called*.

4. B: Choice *B* is correct because the sentence is talking about a continuing process. Therefore, the best modification is to add the word *to* in front of *increase*. Choice *A* is incorrect because this modifier is missing. Choice *C* is incorrect because, with the additional comma, the present tense of *increase* is inappropriate. Choice *D* makes more sense, but the tense is still not the best to use.

5. A: The sentence has no errors, so Choice *A* is correct. Choice *B* is incorrect because it adds an unnecessary comma. Choice *C* is incorrect because *advantage* should not be plural in this sentence unless the singular article *an* is removed. Choice *D* is very tempting. Although this would make the sentence more concise, it leaves out critical information, which makes it incorrect.

Mathematics

The Math section of the SAT focuses on a variety of math concepts and practices including real world applications of math operations and relations. The test includes a total of 58 questions to be answered in 80 minutes. The test contains multiple choice type questions and grid in questions where the test taker will be asked to fill in the answer rather than selecting from a number of options. The test provides instructions on how to complete the grid in questions. Test takers must be careful to only mark one circle in each column of the grid. Answers written only in the boxes above the circles will not be counted. If a fraction is an answer to a grid in question, they need not be simplified, but they must be converted to improper fractions. The Math Test also includes both a calculator section and a no calculator section. Test takers are not allowed to use calculators on the no calculator section; this is to assess math fluency, use of number operations, and number sense.

The Math section will ask the test taker to demonstrate their abilities in problem solving and modeling that reflect the types of situations that students will encounter in future college courses, careers, and everyday life. The Math section is broken down into three major content sections: Heart of Algebra, Problem Solving and Data Analysis, and Passport to Advanced Math. The test also incorporates additional topics in math that will help prepare students for their college and career.

The Heart of Algebra section encompasses the study of algebra and the major concepts needed to solve and create linear equations and functions. Students must also use analytical and problem solving skills to solve questions concerning linear inequalities and systems of equations. Questions in this section are presented in a variety of ways including graphical and algebraic representations of problems that require different strategies and processes to complete.

The Problem Solving and Data Analysis section presents questions involving units and quantities that must be manipulated and converted using ratios, rates, and proportional relationships. Test takers will also need to interpret graphs, charts, and other representations of data to identify key features of a data set such as measures of center, patterns, spread, and deviations.

The Passport to Advanced Math section covers topics that will help ready students for more advanced math topics. One of the major topics in this section is knowledge of expressions and how to work with them. Also included in this section are building functions and interpreting complex equations.

The Additional Topics in Math section covers two main areas: geometric and trigonometric concepts. These include, but are not limited to, volume formulas, trigonometric ratios, and equations and theorems related to circles.

Scores for the Math section of the SAT range from 200 to 800. The three main areas of the Math section, heart of algebra, problem solving and data analysis, and passport to advanced math, are also given subscores which range from 1 to 15.

Heart of Algebra

Solving Linear Equations

A function is called **linear** if it can take the form of the equation $f(x) = ax + b$, or $y = ax + b$, for any two numbers a and b. A linear equation forms a straight line when graphed on the coordinate plane. An example of a linear function is shown below on the graph.

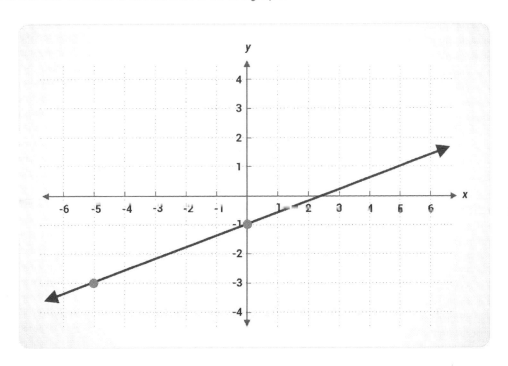

This is a graph of the following function: $y = \frac{2}{5}x - 1$. A table of values that satisfies this function is shown below.

x	y
-5	-3
0	-1
5	1
10	3

These points can be found on the graph using the form (x, y).

When graphing a linear function, note that the ratio of the change of the y coordinate to the change in the x coordinate is constant between any two points on the resulting line, no matter which two points are chosen. In other words, in a pair of points on a line, (x_1, y_1) and (x_2, y_2), with $x_1 \neq x_2$ so that the two points are distinct, then the ratio $\frac{y_2 - y_1}{x_2 - x_1}$ will be the same, regardless of which particular pair of points are chosen. This ratio, $\frac{y_2 - y_1}{x_2 - x_1}$, is called the **slope** of the line and is frequently denoted with the letter m. If slope m is positive, then the line goes upward when moving to the right, while if slope m is negative, then the line goes downward when moving to the right. If the slope is 0, then the line is called **horizontal**, and the *y* coordinate is constant along the entire line. In lines where the *x* coordinate is constant along the entire line, *y* is not actually a function of *x*. For such lines, the slope is not defined. These lines are called **vertical** lines.

Linear functions may take forms other than $y = ax + b$. The most common forms of linear equations are explained below:

1. Standard Form: $Ax + By = C$, in which the slope is given by $m = \frac{-A}{B}$, and the *y*-intercept is given by $\frac{C}{B}$.

2. Slope-Intercept Form: $y = mx + b$, where the slope is *m* and the *y* intercept is *b*.

3. Point-Slope Form: $y - y_1 = m(x - x_1)$, where the slope is *m* and (x_1, y_1) is any point on the chosen line.

4. Two-Point Form: $\frac{y - y_1}{x - x_1} = \frac{y_2 - y_1}{x_2 - x_1}$, where (x_1, y_1) and (x_2, y_2) are any two distinct points on the chosen line. Note that the slope is given by $m = \frac{y_2 - y_1}{x_2 - x_1}$.

5. Intercept Form: $\frac{x}{x_1} + \frac{y}{y_1} = 1$, in which x_1 is the *x*-intercept and y_1 is the *y*-intercept.

These five ways to write linear equations are all useful in different circumstances. Depending on the given information, it may be easier to write one of the forms over another.

If $y = mx$, *y* is directly proportional to *x*. In this case, changing *x* by a factor changes *y* by that same factor. If $y = \frac{m}{x}$, *y* is inversely proportional to *x*. For example, if *x* is increased by a factor of 3, then *y* will be decreased by the same factor, 3.

Sometimes, rather than a situation where there's an equation such as $y = ax + b$ and finding *y* for some value of *x* is requested, the result is given and finding *x* is requested.

The key to solving any equation is to remember that from one true equation, another true equation can be found by adding, subtracting, multiplying, or dividing both sides by the same quantity. In this case, it's necessary to manipulate the equation so that one side only contains *x*. Then the other side will show what *x* is equal to.

For example, in solving $3x - 5 = 2$, adding 5 to each side results in $3x = 7$. Next, dividing both sides by 3 results in $x = \frac{7}{3}$. To ensure the calculated results of *x* is correct, this calculated value can be substituted into the original equation and solved to see if it makes a true statement. For example, $3(\frac{7}{3}) - 5 = 2$ can be simplified by cancelling out the two 3s. This yields $7 - 5 = 2$, which is a true statement.

Sometimes an equation may have more than one *x* term. For example, consider the following equation:

$$3x + 2 = x - 4$$

Moving all of the *x* terms to one side by subtracting *x* from both sides results in $2x + 2 = -4$. Next, subtract 2 from both sides so that there is no constant term on the left side. This yields $2x = -6$. Finally, divide both sides by 2, which leaves $x = -3$.

Solving Linear Inequalities

Solving linear inequalities is very similar to solving equations, except for one rule: when multiplying or dividing an inequality by a negative number, the inequality symbol changes direction. Given the following inequality, solve for x: $-2x + 5 < 13$. The first step in solving this equation is to subtract 5 from both sides. This leaves the inequality: $-2x < 8$. The last step is to divide both sides by -2. By using the rule, the answer to the inequality is $x > -4$.

Since solutions to inequalities include more than one value, number lines are often used to model the answer. For the previous example, the answer is modelled on the number line below. It shows that any number greater than -4, not including -4, satisfies the inequality.

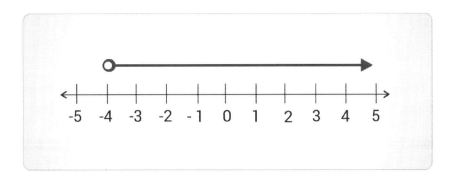

Similar to linear equations, a linear inequality may have a solution set consisting of all real numbers, or can contain no solution. When solved algebraically, a linear inequality in which the variable cancels out and results in a true statement (ex. $7 \geq 2$) has a solution set of all real numbers. A linear inequality in which the variable cancels out and results in a false statement (ex. $7 \leq 2$) has no solution.

Building Linear Functions

A **function** is a special kind of relation where, for each value of x, there is only a single value of y that satisfies the relation. So, $x^2 = y^2$ is *not* a function because in this case, if x is 1, y can be either 1 or -1: the pair (1, 1) and (1, -1) both satisfy the relation. More generally, for this relation, any pair of the form $(a, \pm a)$ will satisfy it. On the other hand, consider the following relation: $y = x^2 + 1$. This is a function because for each value of x, there is a unique value of y that satisfies the relation. Notice, however, there are multiple values of x that give us the same value of y. This is perfectly acceptable for a function. Therefore, y is a function of x.

To determine if a relation is a function, check to see if every x value has a unique corresponding y value.

A function can be viewed as an object that has x as its input and outputs a unique y-value. It is sometimes convenient to express this using **function notation**, where the function itself is given a name, often f. To emphasize that f takes x as its input, the function is written as $f(x)$. In the above example, the equation could be rewritten as $f(x) = x^2 + 1$. To write the value that a function yields for some specific value of x, that value is put in place of x in the function notation. For example, $f(3)$ means the value that the function outputs when the input value is 3. If $f(x) = x^2 + 1$, then $f(3) = 3^2 + 1 = 10$.

A function can also be viewed as a table of pairs (x, y), which lists the value for y for each possible value of x.

The set of all possible values for x in $f(x)$ is called the **domain** of the function, and the set of all possible outputs is called the **range** of the function. Note that usually the domain is assumed to be all real numbers, except those for which the expression for $f(x)$ is not defined, unless the problem specifies otherwise. An example of how a function might not be defined is in the case of $f(x) = \frac{1}{x+1}$, which is not defined when $x = -1$ (which would require dividing by zero). Therefore, in this case the domain would be all real numbers except $x = -1$.

If y is a function of x, then x is the **independent variable** and y is the **dependent variable**. This is because in many cases, the problem will start with some value of x and then see how y changes depending on this starting value.

Functions can be built out of the context of a situation. For example, the relationship between the money paid for a gym membership and the months that someone has been a member can be described through a function. If the one-time membership fee is $40 and the monthly fee is $30, then the function can be written $f(x) = 30x + 40$. The x-value represents the number of months the person has been part of the gym, while the output is the total money paid for the membership. The table below shows this relationship. It is a representation of the function because the initial cost is $40 and the cost increases each month by $30.

x (months)	y (money paid to gym)
0	40
1	70
2	100
3	130

Functions can also be built from existing functions. For example, a given function $f(x)$ can be transformed by adding a constant, multiplying by a constant, or changing the input value by a constant. The new function $g(x) = f(x) + k$ represents a vertical shift of the original function. In $f(x) = 3x - 2$, a vertical shift 4 units up would be:

$$g(x) = 3x - 2 + 4 = 3x + 2$$

Multiplying the function times a constant k represents a vertical stretch, based on whether the constant is greater than or less than 1. The function

$$g(x) = kf(x) = 4(3x - 2) = 12x - 8$$

represents a stretch. Changing the input x by a constant forms the function:

$$g(x) = f(x + k) = 3(x + 4) - 2 = 3x + 12 - 2 = 3x + 10$$

and this represents a horizontal shift to the left 4 units. If $(x - 4)$ was plugged into the function, it would represent a vertical shift.

To evaluate functions, plug in the given value everywhere the variable appears in the expression for the function. For example, find $g(-2)$ where $g(x) = 2x^2 - \frac{4}{x}$. To complete the problem, plug in -2 in the following way:

$$g(-2) = 2(-2)^2 - \frac{4}{-2} \cdot 2 \cdot 4 + 2 = 8 + 2 = 10$$

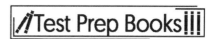

Solving Systems of Inequalities

A **linear inequality in two variables** is a statement expressing an unequal relationship between those two variables. Typically written in slope-intercept form, the variable y can be greater than, less than, greater than or equal to, or less than or equal to a linear expression that includes the variable x, such as $y > 3x$ and $y \leq \frac{1}{2}x - 3$. Questions may include instructions to model real-world scenarios, such as the following:

You work part time cutting lawns for $15 each and cleaning houses for $25 each. Your goal is to make more than $90 this week. Write an inequality to represent the possible pairs of lawns and houses needed to reach your goal.

This scenario can be expressed as $15x + 25y > 90$ where x is the number of lawns cut and y is the number of houses cleaned.

The graph consists of a boundary line dividing the coordinate plane and shading on one side of the boundary. Graph the boundary line just as a linear equation would be graphed. If the inequality symbol is > or <, use a dashed line to indicate that the line is not part of the solution set. If the inequality symbol is > or <, use a solid line to indicate that the boundary line is included in the solution set. Pick an ordered pair (x, y) on either side of the line to test in the inequality statement. If substituting the values for x and y results in a true statement $[15(3) + 25(2) > 90]$, that ordered pair and all others on that side of the boundary line are part of the solution set. To indicate this, shade that region of the graph. If substituting the ordered pair results in a false statement, the ordered pair and all others on that side are not part of the solution set. Therefore, the other region of the graph contains the solutions and should be shaded. The following is an example of the graph of $y \leq x + 2$.

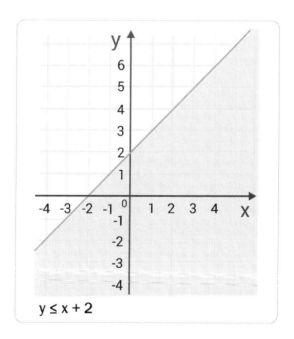

$y \leq x + 2$

Systems of linear inequalities are like systems of equations, but the solutions are different. Since inequalities have infinitely many solutions, their systems also have infinitely many solutions. Finding the solutions of inequalities involves graphs. A system of two equations and two inequalities is linear; thus, the lines can be graphed using slope-intercept form. A system of linear inequalities consists of two linear

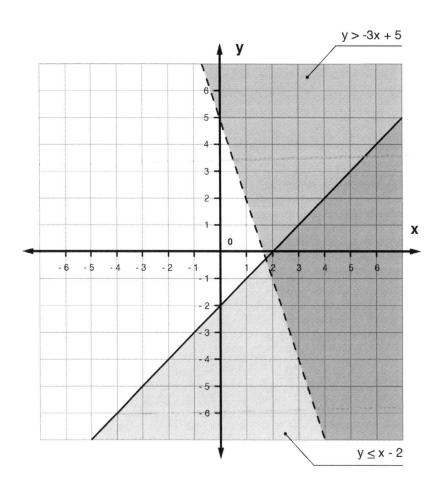
inequalities that make comparisons between two variables. The solution set for a system of inequalities is the region of a graph consisting of ordered pairs that make BOTH inequalities true. To graph the solution set, first graph each linear inequality with appropriate shading. Identify the region of the graph where the shading for the two inequalities overlaps. This region contains the solution set for the system. In the example below, the line with the positive slope is solid, meaning the values on that line are included in the solution. The line with the negative slope is dotted, so the coordinates on that line are not included.

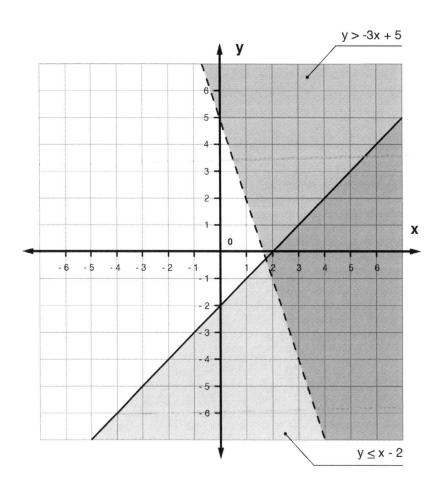

Solving Systems of Linear Equations

A **system of equations** is a group of equations that have the same variables or unknowns. These equations can be linear, but they are not always so. Finding a solution to a system of equations means finding the values of the variables that satisfy each equation. For a linear system of two equations and two variables, there could be a single solution, no solution, or infinitely many solutions.

A single solution occurs when there is one value for x and y that satisfies the system. This would be shown on the graph where the lines cross at exactly one point. When there is no solution, the lines are parallel and do not ever cross. With infinitely many solutions, the equations may look different, but they are the same line. One equation will be a multiple of the other, and on the graph, they lie on top of each other.

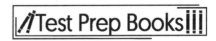

The process of **elimination** can be used to solve a system of equations. For example, the following equations make up a system:

$$x + 3y = 10 \text{ and } 2x - 5y = 9$$

Immediately adding these equations does not eliminate a variable, but it is possible to change the first equation by multiplying the whole equation by -2. This changes the first equation to

$$-2x - 6y = -20$$

The equations can be then added to obtain $-11y = -11$. Solving for y yields $y = 1$. To find the rest of the solution, 1 can be substituted in for y in either original equation to find the value of $x = 7$. The solution to the system is (7, 1) because it makes both equations true, and it is the point in which the lines intersect. If the system is **dependent**—having infinitely many solutions—then both variables will cancel out when the elimination method is used, resulting in an equation that is true for many values of x and y. Since the system is dependent, both equations can be simplified to the same equation or line.

A system can also be solved using **substitution**. This involves solving one equation for a variable and then plugging that solved equation into the other equation in the system. For example, $x - y = -2$ and $3x + 2y = 9$ can be solved using substitution. The first equation can be solved for x, where $x = -2 + y$. Then it can be plugged into the other equation:

$$3(-2 + y) + 2y = 9$$

Solving for y yields:

$$-6 + 3y + 2y = 9$$

That shows that $y = 3$. If $y = 3$, then $x = 1$.

This solution can be checked by plugging in these values for the variables in each equation to see if it makes a true statement.

Finally, a solution to a system of equations can be found graphically. The solution to a linear system is the point or points where the lines cross. The values of x and y represent the coordinates (x, y) where the lines intersect. Using the same system of equations as above, they can be solved for y to put them in slope-intercept form, $y = mx + b$. These equations become $y = x + 2$ and $y = -\frac{3}{2}x + 4.5$. The slope is the coefficient of x, and the y-intercept is the constant value.

This system with the solution is shown below:

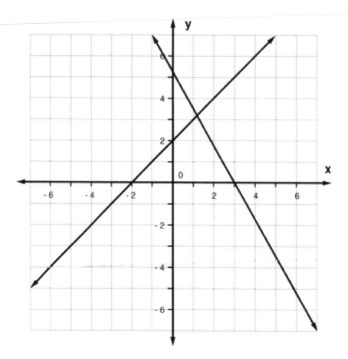

Finding solutions to systems of equations is essentially finding what values of the variables make both equations true. It is finding the input value that yields the same output value in both equations. For functions $g(x)$ and $f(x)$, the equation $g(x) = f(x)$ means the output values are being set equal to each other. Solving for the value of x means finding the x-coordinate that gives the same output in both functions. For example, $f(x) = x + 2$ and $g(x) = -3x + 10$ is a system of equations. Setting $f(x) = g(x)$ yields the equation:

$$x + 2 = -3x + 10$$

Solving for x, gives the x-coordinate $x = 2$ where the two lines cross. This value can also be found by using a table or a graph. On a table, both equations can be given the same inputs, and the outputs can be recorded to find the point(s) where the lines cross. Any method of solving finds the same solution, but some methods are more appropriate for some systems of equations than others.

Interpreting Variables and Constants in Expressions

Algebraic expressions look similar to equations, but they do not include the equal sign. **Algebraic expressions** are comprised of numbers, variables, and mathematical operations. Some examples of algebraic expressions are $8x + 7y - 12z$, $3a^2$, and $5x^3 - 4y^4$.

Algebraic expressions and equations can be used to represent real-life situations and model the behavior of different variables. For example, $2x + 5$ could represent the cost to play games at an arcade. In this case, 5 represents the price of admission to the arcade, and 2 represents the cost of each game played. To calculate the total cost, use the number of games played for x, multiply it by 2, and add 5.

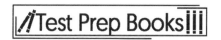

In word problems, multiple quantities are often provided with a request to find some kind of relation between them. This often will mean that one variable (the dependent variable whose value needs to be found) can be written as a function of another variable (the independent variable whose value can be figured from the given information). The usual procedure for solving these problems is to start by giving each quantity in the problem a variable, and then figuring the relationship between these variables.

For example, suppose a car gets 25 miles per gallon. How far will the car travel if it uses 2.4 gallons of fuel? In this case, y would be the distance the car has traveled in miles, and x would be the amount of fuel burned in gallons (2.4). Then the relationship between these variables can be written as an algebraic equation, $y = 25x$. In this case, the equation is $y = 25 \cdot 2.4 = 60$, so the car has traveled 60 miles.

Some word problems require more than just one simple equation to be written and solved. Consider the following situations and the linear equations used to model them.

Suppose Margaret is 2 miles to the east of John at noon. Margaret walks to the east at 3 miles per hour. How far apart will they be at 3 p.m.? To solve this, x would represent the time in hours past noon, and y would represent the distance between Margaret and John. Now, noon corresponds to the equation where x is 0, so the y intercept is going to be 2. It's also known that the slope will be the rate at which the distance is changing, which is 3 miles per hour. This means that the slope will be 3 (be careful at this point: if units were used, other than miles and hours, for x and y variables, a conversion of the given information to the appropriate units would be required first). The simplest way to write an equation given the y-intercept, and the slope is the Slope-Intercept form, which is $y = mx + b$. Recall that m here is the slope, and b is the y intercept. So, $m = 3$ and $b = 2$. Therefore, the equation will be $y = 3x + 2$. The word problem asks how far to the east Margaret will be from John at 3 p.m., which means when x is 3. So, substitute $x = 3$ into this equation to obtain:

$$y = 3 \times 3 + 2 = 9 + 2 = 11$$

Therefore, she will be 11 miles to the east of him at 3 p.m.

For another example, suppose that a box with 4 cans in it weighs 6 lbs., while a box with 8 cans in it weighs 12 lbs. Find out how much a single can weighs. To do this, let x denote the number of cans in the box, and y denote the weight of the box with the cans in lbs. This line touches two pairs: $(4, 6)$ and $(8, 12)$. A formula for this relation could be written using the two-point form, with $x_1 = 4, y_1 = 6, x_2 = 8, y_2 = 12$. This would yield $\frac{y-6}{x-4} = \frac{12-6}{8-4}$, or $\frac{y-6}{x-4} = \frac{6}{4} = \frac{3}{2}$. However, only the slope is needed to solve this problem, since the slope will be the weight of a single can. From the computation, the slope is $\frac{3}{2}$. Therefore, each can weighs $\frac{3}{2}$ lb.

Understanding Connections Between Algebraic and Graphical Representations

To graph relations and functions, the **Cartesian plane** is used. This means to think of the plane as being given a grid of squares, with one direction being the x-axis and the other direction the y-axis. Generally, the independent variable is placed along the horizontal axis, and the dependent variable is placed along the vertical axis. Any point on the plane can be specified by saying how far to go along the x-axis and how far along the y-axis with a pair of numbers (x, y). Specific values for these pairs can be given names such as $C = (-1, 3)$. Negative values mean to move left or down; positive values mean to move right or up. The point where the axes cross one another is called the **origin**. The origin has coordinates $(0, 0)$ and is usually called O when given a specific label.

An illustration of the Cartesian plane, along with the plotted points $(2, 1)$ and $(-1, -1)$, is below.

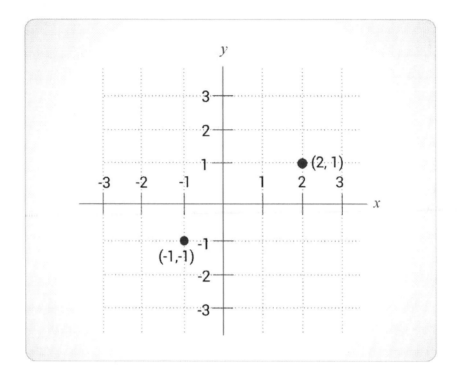

Relations also can be graphed by marking each point whose coordinates satisfy the relation. If the relation is a function, then there is only one value of y for any given value of x. This leads to the **vertical line test**: if a relation is graphed, then the relation is a function if any possible vertical line drawn anywhere along the graph would only touch the graph of the relation in no more than one place. Conversely, when graphing a function, then any possible vertical line drawn will not touch the graph of the function at any point or will touch the function at just one point. This test is made from the definition of a function, where each x-value must be mapped to one and only one y-value.

Equations and inequalities in two variables represent a relationship. Jim owns a car wash and charges $40 per car. The rent for the facility is $350 per month. An equation can be written to relate the number of cars Jim cleans to the money he makes per month. Let x represent the number of cars and y represent the profit Jim makes each month from the car wash. The equation $y = 40x - 350$ can be used to show Jim's profit or loss. Since this equation has two variables, the coordinate plane can be used to show the relationship and predict profit or loss for Jim.

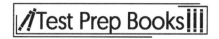

The following graph shows that Jim must wash at least nine cars to pay the rent, where $x = 9$. Anything nine cars and above yield a profit shown in the value on the y-axis.

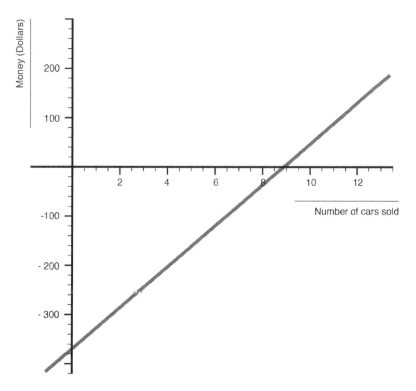

With a single equation in two variables, the solutions are limited only by the situation the equation represents. When two equations or inequalities are used, more constraints are added. For example, in a system of linear equations, there is often—although not always—only one answer. The point of intersection of two lines is the solution. For a system of inequalities, there are infinitely many answers.

Problem Solving and Data Analysis

Ratios, Rates, and Proportions

Ratios are used to show the relationship between two quantities. The ratio of oranges to apples in the grocery store may be 3 to 2. That means that for every 3 oranges, there are 2 apples. This comparison can be expanded to represent the actual number of oranges and apples, such as 36 oranges to 24 apples. Another example may be the number of boys to girls in a math class. If the ratio of boys to girls is given as 2 to 5, that means there are 2 boys to every 5 girls in the class. Ratios can also be compared if the units in each ratio are the same. The ratio of boys to girls in the math class can be compared to the ratio of boys to girls in a science class by stating which ratio is higher and which is lower.

Rates are used to compare two quantities with different units. **Unit rates** are the simplest form of rate. With unit rates, the denominator in the comparison of two units is one. For example, if someone can type at a rate of 1,000 words in 5 minutes, then his or her unit rate for typing is $\frac{1000}{5} = 200$ words in one minute or 200 words per minute. Any rate can be converted into a unit rate by dividing to make the denominator one. 1,000 words in 5 minutes has been converted into the unit rate of 200 words per minute.

Ratios and rates can be used together to convert rates into different units. For example, if someone is driving 50 kilometers per hour, that rate can be converted into miles per hour by using a ratio known as the **conversion factor**. Since the given value contains kilometers and the final answer needs to be in miles, the ratio relating miles to kilometers needs to be used. There are 0.62 miles in 1 kilometer. This, written as a ratio and in fraction form, is

$$\frac{0.62 \ miles}{1 \ km}$$

To convert 50km/hour into miles per hour, the following conversion needs to be set up:

$$\frac{50 \ km}{hour} \times \frac{0.62 \ miles}{1 \ km} = 31 \ miles \ per \ hour$$

The ratio between two similar geometric figures is called the **scale factor**. For example, a problem may depict two similar triangles, A and B. The scale factor from the smaller triangle A to the larger triangle B is given as 2 because the length of the corresponding side of the larger triangle, 16, is twice the corresponding side on the smaller triangle, 8. This scale factor can also be used to find the value of a missing side, x, in triangle A. Since the scale factor from the smaller triangle (A) to larger one (B) is 2, the larger corresponding side in triangle B (given as 25), can be divided by 2 to find the missing side in A ($x = 12.5$). The scale factor can also be represented in the equation $2A = B$ because two times the lengths of A gives the corresponding lengths of B. This is the idea behind similar triangles.

Much like a scale factor can be written using an equation like $2A = B$, a **relationship** is represented by the equation $Y = kX$. X and Y are proportional because as values of X increase, the values of Y also increase. A relationship that is inversely proportional can be represented by the equation $Y = \frac{k}{X}$, where the value of Y decreases as the value of x increases and vice versa.

Proportional reasoning can be used to solve problems involving ratios, percentages, and averages. Ratios can be used in setting up proportions and solving them to find unknowns. For example, if a student completes an average of 10 pages of math homework in 3 nights, how long would it take the student to complete 22 pages? Both ratios can be written as fractions. The second ratio would contain the unknown.

The following proportion represents this problem, where x is the unknown number of nights:

$$\frac{10 \ pages}{3 \ nights} = \frac{22 \ pages}{x \ nights}$$

Solving this proportion entails cross-multiplying and results in the following equation: $10x = 22 \times 3$. Simplifying and solving for x results in the exact solution: $x = 6.6 \ nights$. The result would be rounded up to 7 because the homework would actually be completed on the seventh night.

The following problem uses ratios involving percentages:

If 20% of the class is girls and 30 students are in the class, how many girls are in the class?

To set up this problem, it is helpful to use the common proportion:

$$\frac{\%}{100} = \frac{is}{of}$$

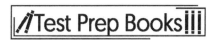

Within the proportion, % is the percentage of girls, 100 is the total percentage of the class, *is* is the number of girls, and *of* is the total number of students in the class. Most percentage problems can be written using this language. To solve this problem, the proportion should be set up as $\frac{20}{100} = \frac{x}{30}$, and then solved for x. Cross-multiplying results in the equation $20 \times 30 = 100x$, which results in the solution $x = 6$. There are 6 girls in the class.

Problems involving volume, length, and other units can also be solved using ratios. A problem may ask for the volume of a cone to be found that has a radius, $r = 7m$ and a height, $h = 16m$. Referring to the formulas provided on the test, the volume of a cone is given as:

$$V = \pi r^2 \frac{h}{3}$$

r is the radius, and h is the height. Plugging $r = 7$ and $h = 16$ into the formula, the following is obtained:

$$V = \pi(7^2)\frac{16}{3}$$

Therefore, volume of the cone is found to be approximately 821m³. Sometimes, answers in different units are sought. If this problem wanted the answer in liters, 821m³ would need to be converted.

Using the equivalence statement 1m³ = 1,000L, the following ratio would be used to solve for liters:

$$821m^3 \times \frac{1000L}{1m^3}$$

Cubic meters in the numerator and denominator cancel each other out, and the answer is converted to 821,000 liters, or $8.21 * 10^5$ L.

Other conversions can also be made between different given and final units. If the temperature in a pool is 30°C, what is the temperature of the pool in degrees Fahrenheit? To convert these units, an equation is used relating Celsius to Fahrenheit. The following equation is used:

$$T_{°F} = 1.8T_{°C} + 32$$

Plugging in the given temperature and solving the equation for T yields the result:

$$T_{°F} = 1.8(30) + 32 = 86°F$$

Both units in the metric system and U.S. customary system are widely used.

Percentages

The word **percent** comes from the Latin phrase for "per one hundred." A percent is a way of writing out a fraction. It is a fraction with a denominator of 100. Thus, $65\% = \frac{65}{100}$.

To convert a fraction to a percent, the denominator is written as 100. For example:

$$\frac{3}{5} = \frac{60}{100} = 60\%$$

In converting a percent to a fraction, the percent is written with a denominator of 100, and the result is simplified. For example:

$$30\% = \frac{30}{100} = \frac{3}{10}$$

The basic percent equation is the following:

$$\frac{is}{of} = \frac{\%}{100}$$

The placement of numbers in the equation depends on what the question asks.

Example 1
Find 40% of 80.

Basically, the problem is asking, "What is 40% of 80?" The 40% is the percent, and 80 is the number to find the percent "of." The equation is:

$$\frac{x}{80} = \frac{40}{100}$$

After cross-multiplying, the problem becomes 100x = 80(40). Solving for x gives the answer: x = 32.

Example 2
What percent of 100 is 20?

The 20 fills in the "is" portion, while 100 fills in the "of." The question asks for the percent, so that will be x, the unknown. The following equation is set up:

$$\frac{20}{100} = \frac{x}{100}$$

Cross-multiplying yields the equation 100x = 20(100). Solving for x gives the answer of 20%.

Example 3
30% of what number is 30?

The following equation uses the clues and numbers in the problem:

$$\frac{30}{x} = \frac{30}{100}$$

Cross-multiplying results in the equation 30(100) = 30x. Solving for x gives the answer x = 100.

Unit Rates and Conversions

When rates are expressed as a quantity of one, they are considered **unit rates**. To determine a unit rate, the first quantity is divided by the second. Knowing a unit rate makes calculations easier than simply having a rate. For example, suppose a 3 pound bag of onions costs $1.77. To calculate the price of 5 pounds of onions, a proportion could show:

$$\frac{3}{1.77} = \frac{5}{x}$$

However, by knowing the unit rate, the value of pounds of onions is multiplied by the unit price. The unit price is calculated:

$$\$1.77/3lb = \$0.59/lb$$

Multiplying the weight of the onions by the unit price yields:

$$5lb \times \frac{\$0.59}{lb} = \$2.95$$

The *lb.* units cancel out.

Similar to unit-rate problems, unit conversions appear in real-world scenarios including cooking, measurement, construction, and currency. Given the conversion rate, unit conversions are written as a fraction (ratio) and multiplied by a quantity in one unit to convert it to the corresponding unit. To determine how many minutes are in $3\frac{1}{2}$ hours, the conversion rate of 60 minutes to 1 hour is written as $\frac{60\ min}{1h}$. Multiplying the quantity by the conversion rate results in:

$$3\frac{1}{2}h \times \frac{60\ min}{1h} = 210\ min$$

(The *h* unit is canceled)

To convert a quantity in minutes to hours, the fraction for the conversion rate is flipped to cancel the *min* unit. To convert 195 minutes to hours, $195min \times \frac{1h}{60\ min}$ is multiplied. The result is $\frac{195h}{60}$ which reduces to $3\frac{1}{4}$h.

Converting units may require more than one multiplication. The key is to set up conversion rates so that units cancel each other out and the desired unit is left. To convert 3.25 yards to inches, given that 1yd = 3ft and 12in = 1ft, the calculation is performed by multiplying:

$$3.25\ yd \times \frac{3ft}{1yd} \times \frac{12in}{1ft}$$

The *yd* and *ft* units will cancel, resulting in 117in.

When working with different systems of measurement, conversion from one unit to another may be necessary. The conversion rate must be known to convert units. One method for converting units is to write and solve a proportion. The arrangement of values in a proportion is extremely important. Suppose that a problem requires converting 20 fluid ounces to cups. To do so, a proportion can be written using the conversion rate of 8fl oz = 1c with x representing the missing value. The proportion can be written in any of the following ways:

$$\frac{1}{8} = \frac{x}{20} \left(\frac{c \; for \; conversion}{fl \; oz \; for \; conversion} = \frac{unknown \; c}{fl \; oz \; given} \right)$$

$$\frac{8}{1} = \frac{20}{x} \left(\frac{fl \; oz \; for \; conversion}{c \; for \; conversion} = \frac{fl \; oz \; given}{unknown \; c} \right)$$

$$\frac{1}{x} = \frac{8}{20} \left(\frac{c \; for \; conversion}{unknown \; c} = \frac{fl \; oz \; for \; conversion}{fl \; oz \; given} \right)$$

$$\frac{x}{1} = \frac{20}{8} \left(\frac{unknown \; c}{c \; for \; conversion} = \frac{fl \; oz \; given}{fl \; oz \; for \; conversion} \right)$$

To solve a proportion, the ratios are cross-multiplied and the resulting equation is solved. When cross-multiplying, all four proportions above will produce the same equation:

$$(8)(x) = (20)(1) \rightarrow 8x = 20$$

Divide by 8 to isolate the variable x, which yields x = 2.5. The variable x represented the unknown number of cups. Therefore, the conclusion is that 20 fluid ounces converts (is equal) to 2.5 cups.

Sometimes converting units requires writing and solving more than one proportion. Suppose an exam question asks to determine how many hours are in 2 weeks. Without knowing the conversion rate between hours and weeks, this can be determined knowing the conversion rates between weeks and days, and between days and hours. First, weeks are converted to days, then days are converted to hours. To convert from weeks to days, the following proportion can be written:

$$\frac{7}{1} = \frac{x}{2} \left(\frac{days \; conversion}{weeks \; conversion} = \frac{days \; unknown}{weeks \; given} \right)$$

Cross-multiplying produces:

$$(7)(2) = (x)(1) \rightarrow 14 = x$$

Therefore, 2 weeks is equal to 14 days. Next, a proportion is written to convert 14 days to hours:

$$\frac{24}{1} = \frac{x}{14} \left(\frac{conversion \; hours}{conversion \; days} = \frac{unknown \; hours}{given \; days} \right)$$

Cross-multiplying produces:

$$(24)(14) = (x)(1) \rightarrow 336 = x$$

Therefore, the answer is that there are 336 hours in 2 weeks.

Scatterplots

A **scatter plot** is a way to visually represent the relationship between two variables. Each variable has its own axis, and usually the independent variable is plotted on the horizontal axis while the dependent variable is plotted on the vertical axis. Data points are plotted in a process that's similar to how ordered pairs are plotted on an *xy*-plane. Once all points from the data set are plotted, the scatter plot is finished. Below is an example of a scatter plot that's plotting the quality and price of an item. Note that price is the independent variable and quality is the dependent variable:

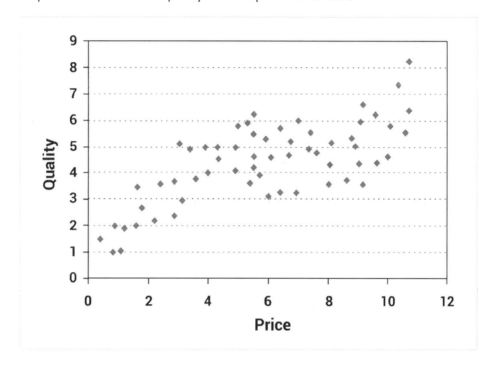

In this example, the quality of the item increases as the price increases.

Regression lines are a way to calculate a relationship between the independent variable and the dependent variable. A straight line means that there's a linear trend in the data. Technology can be used to find the equation of this line (e.g., a graphing calculator or Microsoft Excel®). In either case, all of the data points are entered, and a line is "fit" that best represents the shape of the data. If the line of best-fit has a positive slope (rises from left to right), then the variables have a positive correlation. If the line of best-fit has a negative slope (falls from left to right), then the variables have a negative correlation. If a line of best-fit cannot be drawn, then no correlation exists. A positive or negative correlation can be categorized as strong or weak, depending on how closely the points are graphed around the line of best-fit. Other functions used to model data sets include quadratic and exponential models.

Regression lines can be used to estimate data points not already given. Consider a data set with the average daily temperature at the beach and number of beach visitors. If an equation of a line is found that fits this data set, its input is the average daily temperature and its output is the projected number of visitors. Thus, the number of beach visitors on a 100-degree day can be estimated. The output is a data point on the regression line, and the number of daily visitors is expected to be greater than on a 96-degree day because the regression line has a positive slope.

The formula for a regression line is $y = mx + b$, where m is the slope and b is the *y*-intercept. Both the slope and *y*-intercept are found in the **Method of Least Squares**, which is the process of finding the equation of the line through minimizing residuals. The slope represents the rate of change in *y* as *x* gets larger. Therefore, because *y* is the dependent variable, the slope actually provides the predicted values given the independent variable. The *y*-intercept is the predicted value for when the independent

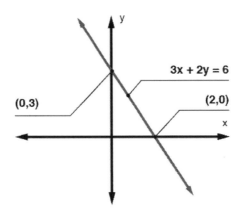

variable equals zero. In the temperature example, the *y*-intercept is the expected number of beach visitors for a very cold average daily temperature of zero degrees.

Investigating Key Features of a Graph

When a linear equation is written in standard form, $Ax + By = C$, it is easy to identify the *x*- and *y*-intercepts for the graph of the line. Just as the *y*-intercept is the point at which the line intercepts the *y*-axis, the *x*-intercept is the point at which the line intercepts the *x*-axis. At the *y*-intercept, $x = 0$, and at the *x*-intercept, $y = 0$. Given an equation in standard form, substitute $x = 0$ to find the *y*-intercept, and substitute $y = 0$ to find the *x*-intercept. For example, to graph $3x + 2y = 6$, substituting 0 for *y* results in $3x + 2(0) = 6$. Solving for *x* yields $x = 2$; therefore, an ordered pair for the line is (2, 0). Substituting 0 for *x* results in $3(0) + 2y = 6$. Solving for *y* yields $y = 3$; therefore, an ordered pair for the line is (0, 3). Plot the two ordered pairs (the *x*- and *y*-intercepts), and construct a straight line through them.

T - chart		Intercepts
x	**y**	x - intercept : (2,0)
0	3	y - intercept : (0,3)
2	0	

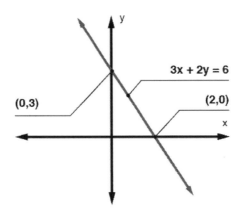

The standard form of a quadratic function is:

$$y = ax^2 + bx + c$$

The graph of a quadratic function is a U-shaped (or upside-down U) curve, called a **parabola**, which is symmetric about a vertical line (axis of symmetry). To graph a parabola, determine its vertex (high or low point for the curve) and at least two points on each side of the axis of symmetry.

Given a quadratic function in standard form, the axis of symmetry for its graph is the line $x = -\frac{b}{2a}$. The vertex for the parabola has an *x*-coordinate of $-\frac{b}{2a}$. To find the *y*-coordinate for the vertex, substitute the calculated *x*-coordinate. To complete the graph, select two different *x*-values, and substitute them into the quadratic function to obtain the corresponding *y*-values. This will give two points on the parabola. Use these two points and the axis of symmetry to determine the two points corresponding to

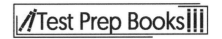

these. The corresponding points are the same distance from the axis of symmetry (on the other side) and contain the same *y*-coordinate.

Plotting the vertex and four other points on the parabola allows for construction of the curve.

Quadratic Function

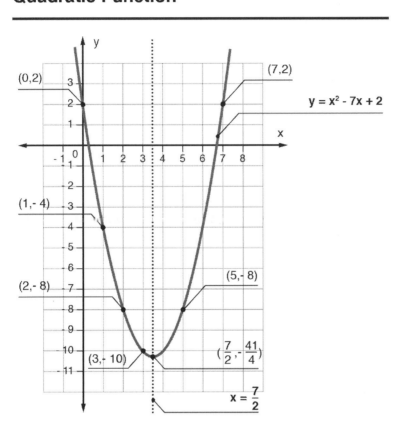

Exponential functions have a general form of $y = (a)(b^x)$. The graph of an exponential function is a curve that slopes upward or downward from left to right. The graph approaches a line, called an **asymptote**, as *x* or *y* increases or decreases. To graph the curve for an exponential function, select *x*-values, and substitute them into the function to obtain the corresponding *y*-values. A general rule of thumb is to select three negative values, zero, and three positive values.

Plotting the seven points on the graph for an exponential function should allow for the construction of a smooth curve through them.

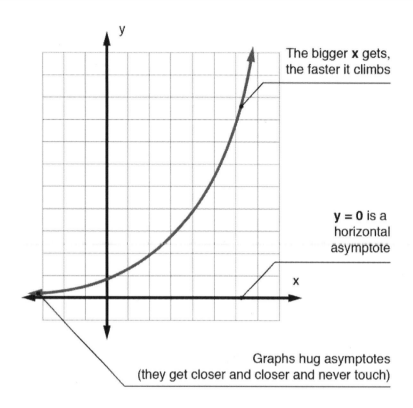

The bigger **x** gets, the faster it climbs

y = 0 is a horizontal asymptote

Graphs hug asymptotes (they get closer and closer and never touch)

Comparing Linear and Exponential Growth

Linear functions are simpler than exponential functions, and the independent variable x has an exponent of 1. Written in the most common form, $y = mx + b$, the coefficient of x indicates how fast the function grows at a constant rate, and the b-value denotes the starting point. An exponential function has an independent variable in the exponent $y = ab^x$. The graph of these types of functions is described as **growth** or **decay**, based on whether the base, b, is greater than or less than 1. These functions are different from quadratic functions because the base stays constant. A common base is base e.

The following two functions model a linear and exponential function respectively: $y = 2x$ and $y = 2^x$. Their graphs are shown below. The first graph, modeling the linear function, shows that the growth is constant over each interval. With a horizontal change of 1, the vertical change is 2. It models a constant positive growth. The second graph models the exponential function, where the horizontal change of 1 yields a vertical change that increases more and more. The exponential graph gets very close to the x-

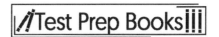

axis, but never touches it, meaning there is an asymptote there. The y-value can never be zero because the base of 2 can never be raised to an input value that yields an output of zero.

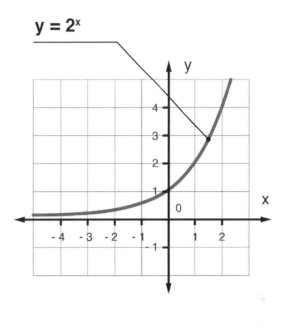

Given a table of values, the type of function can be determined by observing the change in y over equal intervals. For example, the tables below model two functions. The changes in interval for the x-values is 1 for both tables. For the first table, the y-values increase by 5 for each interval. Since the change is constant, the situation can be described as a linear function. The equation would be $y = 5x + 3$. For the second table, the change for y is 5, 20, 100, and 500, respectively. The increases are multiples of 5, meaning the situation can be modeled by an exponential function. The equation $y = 5^x + 3$ models this situation.

x	y
0	3
1	8
2	13
3	18
4	23

x	y
0	3
1	8
2	28
3	128
4	628

Two-Way Tables

Data that isn't described using numbers is known as **categorical data.** For example, age is numerical data but hair color is categorical data. Categorical data is summarized using two-way frequency tables. A **two-way frequency table** counts the relationship between two sets of categorical data. There are rows and columns for each category, and each cell represents frequency information that shows the actual data count between each combination.

For example, the graphic on the left-side below is a two-way frequency table showing the number of girls and boys taking language classes in school. Entries in the middle of the table are known as the **joint frequencies**. For example, the number of girls taking French class is 12, which is a joint frequency. The totals are the **marginal frequencies**. For example, the total number of boys is 20, which is a marginal frequency. If the frequencies are changed into percentages based on totals, the table is known as a **two-way relative frequency table.** Percentages can be calculated using the table total, the row totals, or the column totals.

Here's the process of obtaining the two-way relative frequency table using the table total:

Two-Way Frequency Table

	Spanish	French	German	Total
Boys	10	2	8	20
Girls	15	12	3	30
Total	25	14	11	50

Joint frequencies
Marginal frequencies

Divide all tables entries by the table total (50) →

Two-Way Relative Frequency Table (with respect to table total)

	Spanish	French	German	Total
Boys	0.2	0.04	0.16	0.40
Girls	0.3	0.24	0.06	0.60
Total	0.5	0.28	0.22	1.00

The middle entries are known as **joint probabilities** and the totals are **marginal probabilities**. In this data set, it appears that more girls than boys take Spanish class. However, that might not be the case because more girls than boys were surveyed and the results might be misleading. To avoid such errors, **conditional relative frequencies** are used. The relative frequencies are calculated based on a row or column.

Here are the conditional relative frequencies using column totals:

Two-Way Frequency Table

	Spanish	French	German	Total
Boys	10	2	8	20
Girls	15	12	3	30
Total	25	14	11	50

Divide each column entry by that column's total →

Two-Way Relative Frequency Table (with respect to table total)

	Spanish	French	German	Total
Boys	0.4	0.14	0.73	0.4
Girls	0.6	0.86	0.27	0.6
Total	1.00	1.00	1.00	1.00

Two-way frequency tables can help in making many conclusions about the data. If either the row or column of conditional relative frequencies differs between each row or column of the table, then an association exists between the two categories. For example, in the above tables, the majority of boys are taking German while the majority of girls are taking French. If the frequencies are equal across the rows, there is no association and the variables are labelled as independent. It's important to note that the association does exist in the above scenario, though these results may not occur the next semester when students are surveyed.

When measuring event probabilities, two-way frequency tables can be used to report the raw data and then used to calculate probabilities. If the frequency tables are translated into relative frequency tables, the probabilities presented in the table can be plugged directly into the formulas for conditional probabilities. By plugging in the correct frequencies, the data from the table can be used to determine if events are independent or dependent.

Conditional probability is the probability that event A will happen given that event B has already occurred. An example of this is calculating the probability that a person will eat dessert once they have eaten dinner. This is different than calculating the probability of a person just eating dessert.

The formula for the conditional probability of event A occurring given B is:

$$P(A|B) = \frac{P\ (A \text{ and } B)}{P(B)}$$

It's defined to be the probability of both A and B occurring divided by the probability of event B occurring. If A and B are independent, then the probability of both A and B occurring is equal to $P(A)P(B)$, so $P(A|B)$ reduces to just $P(A)$. This means that A and B have no relationship, and the probability of A occurring is the same as the conditional probability of A occurring given B. Similarly:

$$P(B|A) = \frac{P\ (B \text{ and } A\ )}{P(A)} = P(B)$$

(if A and B are independent)

Two events aren't always independent. For examples, females with glasses and brown hair aren't independent characteristics. There definitely can be overlap because females with brown hair can wear glasses. Also, two events that exist at the same time don't have to have a relationship. For example, even if all females in a given sample are wearing glasses, the characteristics aren't related. In this case, the probability of a brunette wearing glasses is equal to the probability of a female being a brunette multiplied by the probability of a female wearing glasses. This mathematical test of $P(A \cap B) = P(A)P(B)$ verifies that two events are independent.

Conditional probability is the probability that an event occurs given that another event has happened. If the two events are related, the probability that the second event will occur changes if the other event has happened. However, if the two events aren't related and are therefore independent, the first event to occur won't impact the probability of the second event occurring.

Making Inferences About Population Parameters

Inferential statistics attempts to use data about a subset of some population to make inferences about the rest of the population. An example of this would be taking a collection of students who received tutoring and comparing their results to a collection of students who did not receive tutoring, then using that comparison to try to predict whether the tutoring program in question is beneficial.

To be sure that inferences have a high probability of being true for the whole population, the subset that is analyzed needs to resemble a miniature version of the population as closely as possible. For this reason, statisticians like to choose random samples from the population to study, rather than picking a specific group of people based on some similarity. For example, studying the incomes of people who live in Portland does not tell anything useful about the incomes of people who live in Tallahassee.

A **population** is the entire set of people or things of interest. Suppose a study is intended to determine the number of hours of sleep per night for college females in the United States. The population would consist of EVERY college female in the country. A **sample** is a subset of the population that may be used for the study. It would not be practical to survey every female college student, so a sample might consist of one hundred students per school from twenty different colleges in the country. From the results of the survey, a sample statistic can be calculated. A sample statistic is a numerical characteristic of the sample data, including mean and variance. A sample statistic can be used to estimate a corresponding population parameter. A **population parameter** is a numerical characteristic of the entire population. Suppose our sample data had a mean (average) of 5.5. This sample statistic can be used as an estimate of the population parameter (average hours of sleep for every college female in the United States).

A population parameter is usually unknown and therefore estimated using a sample statistic. This estimate may be very accurate or relatively inaccurate based on errors in sampling. A **confidence interval** indicates a range of values likely to include the true population parameter. These are constructed at a given confidence level, such as 95 percent. This means that if the same population is sampled repeatedly, the true population parameter would occur within the interval for 95 percent of the samples.

The accuracy of a population parameter based on a sample statistic may also be affected by **measurement error**. Measurement error is the difference between a quantity's true value and its measured value. Measurement error can be divided into random error and systematic error. An example of random error for the previous scenario would be a student reporting 8 hours of sleep when she sleeps 7 hours per night. Systematic errors are those attributed to the measurement system. Suppose the sleep survey gave response options of 2, 4, 6, 8, or 10 hours. This would lead to systematic measurement error.

Statistics

The field of **statistics** describes relationships between quantities that are related, but not necessarily in a deterministic manner. For example, a graduating student's salary will often be higher when the student graduates with a higher GPA, but this is not always the case. Likewise, people who smoke tobacco are more likely to develop lung cancer, but, in fact, it is possible for non-smokers to develop the disease as well. Statistics describes these kinds of situations, where the likelihood of some outcome depends on the starting data.

Comparing data sets within statistics can mean many things. The first way to compare data sets is by looking at the center and spread of each set. The center of a data set is measured by mean, median, and mode.

Suppose that X is a set of data points $(x_1, x_2, x_3, \ldots x_n)$ and some description of the general properties of this data need to be found.

The first property that can be defined for this set of data is the **mean**. To find the mean, add up all the data points, then divide by the total number of data points. This can be expressed using **summation notation** as:

$$\bar{X} = \frac{x_1 + x_2 + x_3 + \cdots + x_n}{n} = \frac{1}{n}\sum_{i=1}^{n} x_i$$

For example, suppose that in a class of 10 students, the scores on a test were 50, 60, 65, 65, 75, 80, 85, 85, 90, 100. Therefore, the average test score will be:

$$\frac{1}{10}(50 + 60 + 65 + 65 + 75 + 80 + 85 + 85 + 90 + 100) = 75.5$$

The mean is a useful number if the distribution of data is normal (more on this later), which roughly means that the frequency of different outcomes has a single peak and is roughly equally distributed on both sides of that peak. However, it is less useful in some cases where the data might be split or where there are some **outliers**. Outliers are data points that are far from the rest of the data. For example, suppose there are 90 employees and 10 executives at a company. The executives make $1000 per hour, and the employees make $10 per hour. Therefore, the average pay rate will be $\frac{1000 \cdot 10 + 10 \cdot 90}{100} = 109$, or $109 per hour. In this case, this average is not very descriptive.

Another useful measurement is the **median**. In a data set X consisting of data points $x_1, x_2, x_3, \ldots x_n$, the median is the point in the middle. The middle refers to the point where half the data comes before it and half comes after, when the data is recorded in numerical order. If n is odd, then the median is $x_{\frac{n+1}{2}}$. If n is even, it is defined as $\frac{1}{2}\left(x_{\frac{n}{2}} + x_{\frac{n}{2}+1}\right)$, the mean of the two data points closest to the middle of the data points. In the previous example of test scores, the two middle points are 75 and 80. Since there is no single point, the average of these two scores needs to be found. The average is:

$$\frac{75 + 80}{2} = 77.5$$

The median is generally a good value to use if there are a few outliers in the data. It prevents those outliers from affecting the "middle" value as much as when using the mean.

Since an outlier is a data point that is far from most of the other data points in a data set, this means an outlier also is any point that is far from the median of the data set. The outliers can have a substantial effect on the mean of a data set, but they usually do not change the median or mode, or do not change them by a large quantity. For example, consider the data set (3, 5, 6, 6, 6, 8). This has a median of 6 and a mode of 6, with a mean of $\frac{34}{6} \approx 5.67$. Now, suppose a new data point of 1000 is added so that the data set is now (3, 5, 6, 6, 6, 8, 1000). The median and mode, which are both still 6, remain unchanged. However, the average is now $\frac{1034}{7}$, which is approximately 147.7. In this case, the median and mode will be better descriptions for most of the data points.

Outliers in a given data set are sometimes the result of an error by the experimenter, but oftentimes, they are perfectly valid data points that must be taken into consideration.

One additional measure to define for X is the **mode**. This is the data point that appears more frequently. If two or more data points all tie for the most frequent appearance, then each of them is considered a mode. In the case of the test scores, where the numbers were 50, 60, 65, 65, 75, 80, 85, 85, 90, 100, there are two modes: 65 and 85.

The **spread** of a data set refers to how far the data points are from the center (mean or median). The spread can be measured by the range or the quartiles and interquartile range. A data set with all its data points clustered around the center will have a small spread. A data set covering a wide range of values will have a large spread. The **interquartile range** *(IQR)* is the range of the middle 50 percent of the data set. This range can be seen in the large rectangle on a box plot. The **standard deviation** *(σ)* quantifies the amount of variation with respect to the mean. A lower standard deviation shows that the data set doesn't differ greatly from the mean. A larger standard deviation shows that the data set is spread out farther from the mean. The formula used for standard deviation depends on whether it's being used for a population or a sample (a subset of a population). The formula for sample standard deviation is:

$$s = \sqrt{\frac{\Sigma(x - \bar{x})^2}{n - 1}}$$

In this formula, s represents the standard deviation value, x is each value in the data set, $\bar{x}$ is the sample mean, and n is the total number of data points in the set. Note that sample standard deviations use *one less than the total* in the denominator. The population standard deviation formula is similar:

$$\sigma = \sqrt{\frac{\Sigma(x - \mu)^2}{N}}$$

For population standard deviations, sigma (σ) represents the standard deviation, x represents each value in the data set, mu (μ) is the population mean, and N is the total number of data points for the population.

The shape of a data set is another way to compare two or more sets of data. If a data set isn't symmetric around its mean, it's said to be **skewed.** If the tail to the left of the mean is longer, it's said to be **skewed to the left**. In this case, the mean is less than the median. Conversely, if the tail to the right of the mean is longer, it's said to be **skewed to the right** and the mean is greater than the median. When classifying a data set according to its shape, its overall **skewness** is being discussed. If the mean and median are equal, the data set isn't skewed; it is **symmetric**, and is considered normally distributed.

Evaluating Reports

The presentation of statistics can be manipulated to produce a desired outcome. Consider the statement, "Four out of five dentists recommend our toothpaste." This is a vague statement that is obviously biased. (Who are the five dentists this statement references?) This statement is very different from the statement, "Four out of every five dentists recommend our toothpaste." Whether intentional or unintentional, statistics can be misleading. Statistical reports should be examined to verify the validity and significance of the results. The context of the numerical values allows for deciphering the meaning, intent, and significance of the survey or study. Questions on this material will require students to use critical-thinking skills to justify or reject results and conclusions.

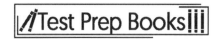

When analyzing a report, consider who conducted the study and their intent. Was it performed by a neutral party or by a person or group with a vested interest? A study on health risks of smoking performed by a health insurance company would have a much different intent than one performed by a cigarette company. Consider the sampling method and the data collection method. Was it a true random sample of the population, or was one subgroup overrepresented or underrepresented?

The three most common types of data gathering techniques are sample surveys, experiments, and observational studies. **Sample surveys** involve collecting data from a random sample of people from a desired population. The measurement of the variable is only performed on this set of people. To have accurate data, the sampling must be unbiased and random. For example, surveying students in an advanced calculus class on how much they enjoy math classes is not a useful sample if the population should be all college students based on the research question. An **experiment** is the method in which a hypothesis is tested using a controlled process called the scientific method.

A cause and the effect of that cause are measured, and the hypothesis is accepted or rejected. Experiments are usually completed in a controlled environment where the results of a control population are compared to the results of a test population. The groups are selected using a randomization process in which each group has a representative mix of the population being tested. Finally, an **observational study** is similar to an experiment. However, this design is used when circumstances prevent or do not allow for a designated control group and experimental group (e.g., lack of funding or unrealistic expectations). Instead, existing control and test populations must be used, so this method has a lack of randomization.

Consider the sleep study scenario from the previous section. If all twenty schools included in the study were state colleges, the results may be biased due to a lack of private-school participants. Consider the measurement system used to obtain the data. Was the system accurate and precise, or was it a flawed system? If, for the sleep study, the possible responses were limited to 2, 4, 6, 8, or 10 hours, it could be argued that the measurement system was flawed. Would odd numbers be rounded up or down? Without clarity of the system, the results could vary greatly. What about students who sleep 12 hours per night? The closest option for them would be 10 hours, which is significantly less.

Every scenario involving statistical reports will be different. The key is to examine all aspects of the study before determining whether to accept or reject the results and corresponding conclusions.

Passport to Advanced Math

Creating Quadratic and Exponential Functions

A polynomial of degree 2 is called **quadratic**. Every quadratic function can be written in the form $ax^2 + bx + c$. The graph of a quadratic function, $y = ax^2 + bx + c$, is called a **parabola**. Parabolas are vaguely U-shaped.

Parabola

Whether the parabola opens upward or downward depends on the sign of a. If a is positive, then the parabola will open upward. If a is negative, then the parabola will open downward. The value of a will also affect how wide the parabola is. If the absolute value of a is large, then the parabola will be fairly skinny. If the absolute value of a is small, then the parabola will be quite wide.

Changes to the value of b affect the parabola in different ways, depending on the sign of a. For positive values of a, increasing b will move the parabola to the left, and decreasing b will move the parabola to the right. On the other hand, if a is negative, the effects will be the opposite: increasing b will move the parabola to the right, while decreasing b will move the parabola to the left.

Changes to the value of c move the parabola vertically. The larger that c is, the higher the parabola gets. This does not depend on the value of a.

The quantity $D = b^2 - 4ac$ is called the **discriminant** of the parabola. When the discriminant is positive, then the parabola has two real zeros, or x intercepts. However, if the discriminant is negative, then there are no real zeros, and the parabola will not cross the x-axis. The highest or lowest point of the parabola is called the **vertex.** If the discriminant is zero, then the parabola's highest or lowest point is on the x-axis, and it will have a single real zero. The x-coordinate of the vertex can be found using the equation $x = -\frac{b}{2a}$. Plug this x-value into the equation and find the y-coordinate.

A quadratic equation is often used to model the path of an object thrown into the air. The x-value can represent the time in the air, while the y-value can represent the height of the object. In this case, the maximum height of the object would be the y-value found when the x-value is $-\frac{b}{2a}$.

An **exponential function** is a function of the form $f(x) = b^x$, where b is a positive real number other than 1. In such a function, b is called the **base**.

The domain of an exponential function is all real numbers, and the range is all positive real numbers. There will always be a horizontal asymptote of $y = 0$ on one side. If b is greater than 1, then the graph will be increasing moving to the right. If b is less than 1, then the graph will be decreasing moving to the right. Exponential functions are one-to-one. The basic exponential function graph will go through the point (0, 1).

Example

Solve $5^{x+1} = 25$.

Get the x out of the exponent by rewriting the equation $5^{x+1} = 5^2$ so that both sides have a base of 5.

Since the bases are the same, the exponents must be equal to each other.

This leaves $x + 1 = 2$ or $x = 1$.

To check the answer, the x-value of 1 can be substituted back into the original equation.

Determining Forms of Expressions

There is a four-step process in problem-solving that can be used as a guide:

1. Understand the problem and determine the unknown information.

2. Translate the verbal problem into an algebraic equation.

3. Solve the equation by using inverse operations.

4. Check the work and answer the given question.

Example

Three times the sum of a number plus 4 equals the number plus 8. What is the number?

The first step is to determine the unknown, which is the number, or x.

The second step is to translate the problem into the equation, which is $3(x + 4) = x + 8$.

The equation can be solved as follows:

$3x + 12 = x + 8$	Apply the distributive property
$3x = x - 4$	Subtract 12 from both sides of the equation
$2x = -4$	Subtract x from both sides of the equation
$x = -2$	Divide both sides of the equation by 2

The final step is checking the solution. Plugging the value for x back into the equation yields the following problem:

$$3(-2) + 12 = -2 + 8$$

Using the order of operations shows that a true statement is made: $6 = 6$

The four-step process of problem solving can be used with geometric reasoning problems as well. There are many geometric properties and terminology included within geometric reasoning.

For example, the perimeter of a rectangle can be written in the terms of the width, or the width can be written in terms of the length.

Example

The width of a rectangle is 2 centimeters less than the length. If the perimeter of the rectangle is 44 centimeters, then what are the dimensions of a rectangle?

The first step is to determine the unknown, which is in terms of the length, l.

The second step is to translate the problem into the equation using the perimeter of a rectangle, $P = 2l + 2w$. The width is the length minus 2 centimeters. The resulting equation is:

$$2l + 2(l - 2) = 44$$

The equation can be solved as follows:

$2l + 2l - 4 = 44$	Apply the distributive property on the left side of the equation
$4l - 4 = 44$	Combine like terms on the left side of the equation
$4l = 48$	Add 4 to both sides of the equation
$l = 12$	Divide both sides of the equation by 4

The length of the rectangle is 12 centimeters. The width is the length minus 2 centimeters, which is 10 centimeters. Checking the answers for length and width forms the following equation:

$$44 = 2(12) + 2(10)$$

The equation can be solved using the order of operations to form a true statement: $44 = 44$.

Equations can also be created from complementary angles (angles that add up to 90°) and supplementary angles (angles that add up to 180°).

Example

Two angles are complementary. If one angle is four times the other angle, what is the measure of each angle?

The first step is to determine the unknown, which is the measure of the first angle. (It could be the measure of either, but we will choose to solve for the first angle.)

The second step is to translate the problem into the equation using the known statement: the sum of two complementary angles is 90°. The resulting equation is $4x + x = 90$. The equation can be solved as follows:

$5x = 90$	Combine like terms on the left side of the equation
$x = 18$	Divide both sides of the equation by 5

The first angle is 18° and the second angle is 4 times the unknown, which is 4 times 18 or 72°.

Going back to check the answer with the original question, 72 and 18 have a sum of 90, making them complementary angles. Seventy-two degrees is also four times the other angle, 18 degrees.

Creating Equivalent Expressions Involving Exponents and Radicals

An **exponent** is written as a^b. In this expression, a is called the **base** and b is called the **exponent**. It is properly stated that a is raised to the n-th power. Therefore, in the expression 2^3, the exponent is 3, while the base is 2. Such an expression is called an **exponential expression**. Note that when the exponent is 2, it is called **squaring** the base, and when it is 3, it is called **cubing** the base.

When the exponent is a positive integer, this indicates the base is multiplied by itself the number of times written in the exponent. So, in the expression 2^3, multiply 2 by itself with 3 copies of 2:

$$2^3 = 2 \times 2 \times 2 = 8$$

One thing to notice is that, for positive integers n and m, $a^n a^m = a^{n+m}$ is a rule. In order to make this rule be true for an integer of 0, $a^0 = 1$, so that:

$$a^n a^0 = a^{n+0} = a^n$$

And, in order to make this rule be true for negative exponents, $a^{-n} = \frac{1}{a^n}$.

Another rule for simplifying expressions with exponents is shown by the following equation: $(a^m)^n = a^{mn}$. This is true for fractional exponents as well. So, for a positive integer, define $a^{\frac{1}{n}}$ to be the number that, when raised to the n-th power, provides a. In other words, $(a^{\frac{1}{n}})^n = a$ is the desired equation. It should be noted that $a^{\frac{1}{n}}$ is the n-th root of a. This also can be written as $a^{\frac{1}{n}} = \sqrt[n]{a}$. The symbol on the right-hand side of this equation is called a **radical.** If the root is left out, assume that the 2nd root should be taken, also called the **square** root:

$$a^{\frac{1}{2}} = \sqrt[2]{a} = \sqrt{a}$$

Additionally, $\sqrt[3]{a}$ is also called the **cube** root.

Note that when multiple roots exist, $a^{\frac{1}{n}}$ is defined to be the **positive** root. So, $4^{\frac{1}{2}} = 2$. Also note that negative numbers do not have even roots in the real numbers.

This also enables finding exponents for any rational number:

$$a^{\frac{m}{n}} = (a^{\frac{1}{n}})^m = (a^m)^{\frac{1}{n}}$$

In fact, the exponent can be any real number. In general, the following rules for exponents should be used for any numbers $a, b, m,$ and n.

- $a^1 = a$.
- $1^a = 1$.
- $a^0 = 1$.
- $a^m a^n = a^{m+n}$.
- $\dfrac{a^m}{a^n} = a^{m-n}$

- $(a^m)^n = a^{m \times n}$.
- $(ab)^m = a^m b^m$.
- $\left(\frac{a}{b}\right)^m = \frac{a^m}{b^m}$.

As an example of applying these rules, consider the problem of simplifying the expression:

$$(3x^2 y)^3 (2xy^4)$$

Start by simplifying the left term using the sixth rule listed. Applying this rule yields the following expression

$$27x^6 y^3 (2xy^4)$$

The exponents can now be combined with base x and the exponents with base y. Multiply the coefficients to yield $54x^7 y^7$.

In mathematical expressions containing exponents and other operations, the order of operations must be followed. PEMDAS states that exponents are calculated after any parenthesis and grouping symbols but before any multiplication, division, addition, and subtraction.

Here are some of the most important properties of exponents and roots: if n is an integer, and if $a^n = b^n$, then $a = b$ if n is odd; but $a = \pm b$ if n is even. Similarly, if the roots of two things are equal, $\sqrt[n]{a} = \sqrt[n]{b}$, then $a = b$. This means that when starting with a true equation, both sides of that equation can be raised to a given power to obtain another true equation. Beware that when an even-powered root is taken on both sides of the equation, a $\pm$ in the result. For example, given the equation $x^2 = 16$, take the square root of both sides to solve for x. This results in the answer $x = \pm 4$ because $(-4)^2 = 16$ and $(4)^2 = 16$.

Another property is that if $a^n = a^m$, then $n = m$. This is true for any real numbers n and m.

For solving the equation $\sqrt{x + 2} - 1 = 3$, start by moving the -1 over to the right-hand side. This is performed by adding 1 to both sides, which yields $\sqrt{x + 2} = 4$. Now, square both sides, but remember that by squaring both sides, the signs are irrelevant. This yields $x + 2 = 16$, which simplifies to give $x = 14$.

Now consider the problem $(x + 1)^4 = 16$. To solve this, take the 4th root of both sides, which means an ambiguity in the sign will be introduced because it is an even root:

$$\sqrt[4]{(x + 1)^4} = \pm\sqrt[4]{16}$$

The right-hand side is 2, since $2^4 = 16$. Therefore:

$$x + 1 = \pm 2 \text{ or } x = -1 \pm 2$$

Thus, the two possible solutions are $x = -3$ and $x = 1$.

Remember that when solving equations, the answer can be checked by plugging the solution back into the problem to make a true statement.

Creating Equivalent Forms of Expressions

Algebraic expressions are made up of numbers, variables, and combinations of the two, using mathematical operations. Expressions can be rewritten based on their factors. For example, the expression $6x + 4$ can be rewritten as $2(3x + 2)$ because 2 is a factor of both $6x$ and 4. More complex expressions can also be rewritten based on their factors. The expression $x^4 - 16$ can be rewritten as $(x^2 - 4)(x^2 + 4)$. This is a different type of factoring, where a difference of squares is factored into a sum and difference of the same two terms. With some expressions, the factoring process is simple and only leads to a different way to represent the expression. With others, factoring and rewriting the expression leads to more information about the given problem.

In the following quadratic equation, factoring the binomial leads to finding the zeros of the function:

$$x^2 - 5x + 6 = y$$

This equations factors into $(x - 3)(x - 2) = y$, where 2 and 3 are found to be the zeros of the function when y is set equal to zero. The zeros of any function are the x-values where the graph of the function on the coordinate plane crosses the x-axis.

Factoring an equation is a simple way to rewrite the equation and find the zeros, but factoring is not possible for every quadratic. Completing the square is one way to find zeros when factoring is not an option. The following equation cannot be factored: $x^2 + 10x - 9 = 0$. The first step in this method is to move the constant to the right side of the equation, making it $x^2 + 10x = 9$. Then, the coefficient of x is divided by 2 and squared. This number is then added to both sides of the equation, to make the equation still true. For this example, $\left(\frac{10}{2}\right)^2 = 25$ is added to both sides of the equation to obtain:

$$x^2 + 10x + 25 = 9 + 25$$

This expression simplifies to $x^2 + 10x + 25 = 34$, which can then be factored into:

$$(x + 5)^2 = 34$$

Solving for x then involves taking the square root of both sides and subtracting 5. This leads to two zeros of the function:

$$x = \pm\sqrt{34} - 5$$

Depending on the type of answer the question seeks, a calculator may be used to find exact numbers.

Given a quadratic equation in standard form— $ax^2 + bx + c = 0$—the sign of a tells whether the function has a minimum value or a maximum value. If $a > 0$, the graph opens up and has a minimum value. If $a < 0$, the graph opens down and has a maximum value. Depending on the way the quadratic equation is written, multiplication may need to occur before a max/min value is determined.

There are also properties of numbers that are true for certain operations. The **commutative** property allows the order of the terms in an expression to change while keeping the same final answer. Both addition and multiplication can be completed in any order and still obtain the same result. However, order does matter in subtraction and division. The **associative** property allows any terms to be "associated" by parenthesis and retain the same final answer. For example:

$$(4 + 3) + 5 = 4 + (3 + 5)$$

Both addition and multiplication are associative; however, subtraction and division do not hold this property. The **distributive** property states that $a(b + c) = ab + ac$. It is a property that involves both addition and multiplication, and the a is distributed onto each term inside the parentheses.

The expression $4(3 + 2)$ is simplified using the order of operations. Simplifying inside the parenthesis first produces 4×5, which equals 20. The expression $4(3 + 2)$ can also be simplified using the distributive property:

$$4(3 + 2) = 4 \times 3 + 4 \times 2 = 12 + 8 = 20$$

Consider the following example: $4(3x - 2)$. The expression cannot be simplified inside the parenthesis because $3x$ and -2 are not like terms and therefore cannot be combined. However, the expression can be simplified by using the distributive property and multiplying each term inside of the parenthesis by the term outside of the parenthesis: $12x - 8$. The resulting equivalent expression contains no like terms, so it cannot be further simplified.

Consider the expression:

$$(3x + 2y + 1) - (5x - 3) + 2(3y + 4)$$

Again, there are no like terms, but the distributive property is used to simplify the expression. Note there is an implied one in front of the first set of parentheses and an implied -1 in front of the second set of parentheses. Distributing the 1, -1, and 2 produces:

$$1(3x) + 1(2y) + 1(1) - 1(5x) - 1(-3) + 2(3y) + 2(4)$$

$$3x + 2y + 1 - 5x + 3 + 6y + 8$$

This expression contains like terms that are combined to produce the simplified expression:

$$-2x + 8y + 12$$

Algebraic expressions are tested to be equivalent by choosing values for the variables and evaluating both expressions. For example, $4(3x - 2)$ and $12x - 8$ are tested by substituting 3 for the variable x and calculating to determine if equivalent values result.

Solving Quadratic Equations

A **quadratic equation** is an equation in the form:

$$ax^2 + bx + c = 0$$

There are several methods to solve such equations. The easiest method will depend on the quadratic equation in question.

Sometimes, it is possible to solve quadratic equations by manually **factoring** them. This means rewriting them in the form $(x + A)(x + B) = 0$. If this is done, then they can be solved by remembering that when $ab = 0$, either a or b must be equal to zero. Therefore, to have $(x + A)(x + B) = 0$, $(x + A) = 0$ or $(x + B) = 0$ is needed. These equations have the solutions $x = -A$ and $x = -B$, respectively.

In order to factor a quadratic equation, note that:

$$(x + A)(x + B) = x^2 + (A + B)x + AB$$

So, if an equation is in the form $x^2 + bx + c$, two numbers, A and B, need to be found that will add up to give us b, and multiply together to give us c.

As an example, consider solving the equation:

$$-3x^2 + 6x + 9 = 0$$

Start by dividing both sides by -3, leaving:

$$x^2 - 2x - 3 = 0$$

Now, notice that $1 - 3 = -2$, and also that:

$$(1)(-3) = -3$$

This means the equation can be factored into:

$$(x + 1)(x - 3) - 0$$

Now, solve $(x + 1) = 0$ and $(x - 3) = 0$ to get $x = -1$ and $x = 3$ as the solutions.

It is useful when trying to factor to remember that:

$$x^2 + 2xy + y^2 = (x + y)^2$$
$$x^2 - 2xy + y^2 = (x - y)^2$$
$$x^2 - y^2 = (x + y)(x - y)$$

However, factoring by hand is often hard to do. If there are no obvious ways to factor the quadratic equation, solutions can still be found by using the **quadratic formula**.

The quadratic formula is:

$$x = \frac{-b \pm \sqrt{b^2 - 4ac}}{2a}$$

This method will always work, although it sometimes can take longer than factoring by hand, if the factors are easy to guess. Using the standard form $ax^2 + bx + c = 0$, plug the values of a, b, and c from the equation into the formula and solve for x. There will either be two answers, one answer, or no real answer. No real answer comes when the value of the discriminant, the number under the square root, is a negative number. Since there are no real numbers that square to get a negative, the answer will be no real roots.

Here is an example of solving a quadratic equation using the quadratic formula. Suppose the equation to solve is:

$$-2x^2 + 3x + 1 = 0$$

There is no obvious way to factor this, so the quadratic formula is used, with $a = -2, b = 3, c = 1$. After substituting these values into the quadratic formula, it yields this:

$$x = \frac{-3 \pm \sqrt{3^2 - 4(-2)(1)}}{2(-2)}$$

This can be simplified to obtain:

$$\frac{3 \pm \sqrt{9 + 8}}{4}$$

or

$$\frac{3 \pm \sqrt{17}}{4}$$

Challenges can be encountered when asked to find a quadratic equation with specific roots. Given roots A and B, a quadratic function can be constructed with those roots by taking $(x - A)(x - B)$. So, in constructing a quadratic equation with roots $x = -2, 3$, it would result in:

$$(x + 2)(x - 3) = x^2 - x - 6$$

Multiplying this by a constant also could be done without changing the roots.

Adding, Subtracting, and Multiplying Polynomial Expressions

An expression of the form ax^n, where n is a non-negative integer, is called a **monomial** because it contains one term. A sum of monomials is called a **polynomial.** For example, $-4x^3 + x$ is a polynomial, while $5x^7$ is a monomial. A function equal to a polynomial is called a **polynomial function**.

The monomials in a polynomial are also called the **terms** of the polynomial.

The constants that precede the variables are called **coefficients**.

The highest value of the exponent of x in a polynomial is called the **degree** of the polynomial. So, $-4x^3 + x$ has a degree of 3, while $-2x^5 + x^3 + 4x + 1$ has a degree of 5. When multiplying polynomials, the degree of the result will be the sum of the degrees of the two polynomials being multiplied.

To add polynomials, add the coefficients of like powers of x. For example:

$$(-2x^5 + x^3 + 4x + 1) + (-4x^3 + x)$$

$$-2x^5 + (1 - 4)x^3 + (4 + 1)x + 1$$

$$-2x^5 - 3x^3 + 5x + 1$$

Likewise, subtraction of polynomials is performed by subtracting coefficients of like powers of x. So:

$$(-2x^5 + x^3 + 4x + 1) - (-4x^3 + x)$$

$$-2x^5 + (1 + 4)x^3 + (4 - 1)x + 1$$

$$-2x^5 + 5x^3 + 3x + 1$$

To multiply two polynomials, multiply each term of the first polynomial by each term of the second polynomial and add the results. For example:

$$(4x^2 + x)(-x^3 + x)$$

$$4x^2(-x^3) + 4x^2(x) + x(-x^3) + x(x)$$

$$-4x^5 + 4x^3 - x^4 + x^2$$

In the case where each polynomial has two terms, like in this example, some students find it helpful to remember this as multiplying the First terms, then the Outer terms, then the Inner terms, and finally the Last terms, with the mnemonic FOIL. For longer polynomials, the multiplication process is the same, but there will be, of course, more terms, and there is no common mnemonic to remember each combination.

Solving Equations with Radicals or Variables in the Denominator

When solving radical and rational equations, extraneous solutions must be accounted for when finding the answers. For example, the equation $\frac{x}{x-5} - \frac{3x}{x+3}$ has two values that create a 0 denominator: $x \neq 5, -3$. When solving for x, these values must be considered because they cannot be solutions. In the given equation, solving for x can be done using cross-multiplication, yielding the equation:

$$x(x + 3) = 3x(x - 5)$$

Distributing results in the quadratic equation $x^2 + 3x = 3x^2 - 15x$; therefore, all terms must be moved to one side of the equals sign. This results in $2x^2 - 18x = 0$, which in factored form is $2x(x - 9) = 0$. Setting each factor equal to zero, the apparent solutions are $x = 0$ and $x = 9$. These two solutions are neither 5 nor -3, so they are viable solutions. Neither 0 nor 9 create a 0 denominator in the original equation.

A similar process exists when solving radical equations. One must check to make sure the solutions are defined in the original equations. Solving an equation containing a square root involves isolating the root and then squaring both sides of the equals sign. Solving a cube root equation involves isolating the radical and then cubing both sides. In either case, the variable can then be solved for because there are no longer radicals in the equation.

For example, the following expression is a radical that can be simplified: $\sqrt{24x^2}$. First, the number must be factored out to the highest perfect square. Any perfect square can be taken out of a radical. Twenty-four can be factored into 4 and 6, and 4 can be taken out of the radical. $\sqrt{4} = 2$ can be taken out, and 6 stays underneath. If $x > 0$, x can be taken out of the radical because it is a perfect square. The simplified radical is $2x\sqrt{6}$. An approximation can be found using a calculator.

Solving a System with One Linear Equation and One Quadratic Equation

A system of equations may also be made up of a linear and a quadratic equation. These systems may have one solution, two solutions, or no solutions. The graph of these systems involves one straight line and one parabola. Algebraically, these systems can be solved by solving the linear equation for one variable and plugging that answer in to the quadratic equation. If possible, the equation can then be solved to find part of the answer. The graphing method is commonly used for these types of systems. On a graph, these two lines can be found to intersect at one point, at two points across the parabola, or at no points.

Solving a system of one linear equation and one quadratic equation algebraically involves using the substitution method. Consider the following system: $y = x^2 + 9x + 11$; $y = 2x - 1$. Substitute the equivalent value of y from the linear equation $(2x - 1)$ into the quadratic equation. The resulting equation would be:

$$2x - 1 = x^2 + 9x + 11$$

Next, solve the resulting quadratic equation using the appropriate method—factoring, taking square roots, or using the quadratic formula. This equation can be solved by factoring:

$$0 = x^2 + 7x + 120$$

$$(x + 3)(x + 4)$$

$$x + 3 = 0 \text{ or } x + 4 = 0$$

$$x = -3 \text{ or } x = -4$$

Next, find the corresponding y-values by substituting the x-values into the original linear equation:

$$y = 2(-4) - 1 = -9$$

$$y = 2(-3) - 1 = -7$$

Write the solutions as ordered pairs: $(-4, -9)$ and $(-3, -7)$. Finally, check the possible solutions by substituting each into both original equations. (In this case, both solutions "check out.")

Rewriting Simple Rational Expressions

A fraction, or ratio, wherein each part is a polynomial, defines **rational expressions**. Some examples include $\frac{2x+6}{x}$, $\frac{1}{x^2-4x+8}$, and $\frac{z^2}{x+5}$. Exponents on the variables are restricted to whole numbers, which means roots and negative exponents are not included in rational expressions.

Rational expressions can be transformed by factoring. For example, the expression $\frac{x^2-5x+6}{(x-3)}$ can be rewritten by factoring the numerator to obtain:

$$\frac{(x - 3)(x - 2)}{(x - 3)}$$

Therefore, the common binomial $(x - 3)$ can cancel so that the simplified expression is:

$$\frac{(x - 2)}{1} = (x - 2)$$

Additionally, other rational expressions can be rewritten to take on different forms. Some may be factorable in themselves, while others can be transformed through arithmetic operations. Rational expressions are closed under addition, subtraction, multiplication, and division by a nonzero expression.

Closed means that if any one of these operations is performed on a rational expression, the result will still be a rational expression. The set of all real numbers is another example of a set closed under all four operations.

Adding and subtracting rational expressions is based on the same concepts as adding and subtracting simple fractions. For both concepts, the denominators must be the same for the operation to take place. For example, here are two rational expressions:

$$\frac{x^3 - 4}{(x - 3)} + \frac{x + 8}{(x - 3)}$$

Since the denominators are both $(x - 3)$, the numerators can be combined by collecting like terms to form:

$$\frac{x^3 + x + 4}{(x - 3)}$$

If the denominators are different, they need to be made common (the same) by using the **Least Common Denominator** (LCD). Each denominator needs to be factored, and the LCD contains each factor that appears in any one denominator the greatest number of times it appears in any denominator. The original expressions need to be multiplied times a form of 1, which will turn each denominator into the LCD. This process is like adding fractions with unlike denominators. It is also important when working with rational expressions to define what value of the variable makes the denominator zero. For this particular value, the expression is undefined.

Multiplication of rational expressions is performed like multiplication of fractions. The numerators are multiplied; then, the denominators are multiplied. The final fraction is then simplified. The expressions are simplified by factoring and cancelling out common terms. In the following example, the numerator of the second expression can be factored first to simplify the expression before multiplying:

$$\frac{x^2}{(x - 4)} \times \frac{x^2 - x - 12}{2}$$

$$\frac{x^2}{(x - 4)} \times \frac{(x - 4)(x + 3)}{2}$$

The $(x - 4)$ on the top and bottom cancel out:

$$\frac{x^2}{1} \times \frac{(x + 3)}{2}$$

Then multiplication is performed, resulting in:

$$\frac{x^3 + 3x^2}{2}$$

Dividing rational expressions is similar to the division of fractions, where division turns into multiplying by a reciprocal. So the following expression can be rewritten as a multiplication problem:

$$\frac{x^2 - 3x + 7}{x - 4} \div \frac{x^2 - 5x + 3}{x - 4}$$

$$\frac{x^2 - 3x + 7}{x - 4} \times \frac{x - 4}{x^2 - 5x + 3}$$

The $x - 4$ cancels out, leaving:

$$\frac{x^2 - 3x + 7}{x^2 - 5x + 3}$$

The final answers should always be completely simplified. If a function is composed of a rational expression, the zeros of the graph can be found from setting the polynomial in the numerator as equal to zero and solving. The values that make the denominator equal to zero will either exist on the graph as a hole or a vertical asymptote.

Interpreting Parts of Nonlinear Expressions

When a nonlinear function is used to model a real-life scenario, some aspects of the function may be relevant, while others may not. The context of each scenario will dictate what should be used. In general, x- and y-intercepts will be points of interest. A y-intercept is the value of y when $x = 0$, and an x-intercept is the value of x when $y = 0$. Suppose a nonlinear function models the value of an investment (y) over the course of time (x). It would be relevant to determine the initial value. This initial value would be the y-intercept of the function (where time $= 0$). It would also be useful to note any point in time in which the value of the investment would be 0. These would be the x-intercepts of the function.

Another aspect of a function that is typically desired is the **rate of change**. This tells how fast the outputs are growing or decaying with respect to given inputs. The rate of change for a quadratic function in standard form, $y = ax^2 + bx + c$, is determined by the value of a. A positive value indicates growth, and a negative value indicates decay. The rate of change for an exponential function in standard form, $y = (a)(b^x)$, is determined by the value of b. If b is greater than 1, the function describes exponential growth, and if b is less than 1, the function describes exponential decay.

For polynomial functions, the rate of change can be estimated by the highest power of the function. Polynomial functions also include absolute and/or relative minimums and maximums. Consider functions modeling production or expenses. Maximum and minimum values would be relevant aspects of these models.

Finally, the domain and range for a function should be considered for relevance. The domain consists of all input values, and the range consists of all output values. Suppose a function models the volume of a container to be produced in relation to its height. Although the function that models the scenario may include negative values for inputs and outputs, these parts of the function would obviously not be relevant.

Understanding the Relationship Between Zeros and Factors of Polynomials

Finding the zeros of polynomial functions is the same process as finding the solutions of polynomial equations. These are the points at which the graph of the function crosses the x-axis. As stated previously, factors can be used to find the zeros of a polynomial function. The degree of the function shows the number of possible zeros. If the highest exponent on the independent variable is 4, then the degree is 4, and the number of possible zeros is 4. If there are complex solutions, the number of roots is less than the degree.

Given the function $y = x^2 + 7x + 6$, y can be set equal to zero, and the polynomial can be factored. The equation turns into $0 = (x + 1)(x + 6)$, where $x = -1$ and $x = -6$ are the zeros. Since this is a quadratic equation, the shape of the graph will be a parabola. Knowing that zeros represent the points where the parabola crosses the x-axis, the maximum or minimum point is the only other piece needed to sketch a rough graph of the function. By looking at the function in standard form, the coefficient of x is positive; therefore, the parabola opens *up*.

Using the zeros and the minimum, the following rough sketch of the graph can be constructed:

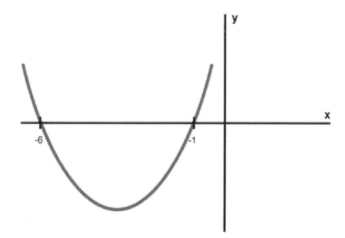

Factors for polynomials are similar to factors for integers—they are numbers, variables, or polynomials that, when multiplied together, give a product equal to the polynomial in question. One polynomial is a factor of a second polynomial if the second polynomial can be obtained from the first by multiplying by a third polynomial.

$6x^6 + 13x^4 + 6x^2$ can be obtained by multiplying together $(3x^4 + 2x^2)(2x^2 + 3)$. This means $2x^2 + 3$ and $3x^4 + 2x^2$ are factors of $6x^6 + 13x^4 + 6x^2$.

In general, finding the factors of a polynomial can be tricky. However, there are a few types of polynomials that can be factored in a straightforward way.

If a certain monomial is in each term of a polynomial, it can be factored out. There are several common forms polynomials take, which if you recognize, you can solve. The first example is a perfect square trinomial. To factor this polynomial, first expand the middle term of the expression:

$$x^2 + 2xy + y^2$$

$$x^2 + xy + xy + y^2$$

Factor out a common term in each half of the expression (in this case x from the left and y from the right):

$$x(x + y) + y(x + y)$$

Then the same can be done again, treating $(x + y)$ as the common factor:

$$(x + y)(x + y) = (x + y)^2$$

Therefore, the formula for this polynomial is:

$$x^2 + 2xy + y^2 = (x + y)^2$$

Next is another example of a perfect square trinomial. The process is the similar, but notice the difference in sign:

$$x^2 - 2xy + y^2$$

$$x^2 - xy - xy + y^2$$

Factor out the common term on each side:

$$x(x - y) - y(x - y)$$

Factoring out the common term again:

$$(x - y)(x - y) = (x - y)^2$$

Thus:

$$x^2 - 2xy + y^2 = (x - y)^2$$

The next is known as a difference of squares. This process is effectively the reverse of binomial multiplication:

$$x^2 - y^2$$

$$x^2 - xy + xy - y^2$$

$$x(x - y) + y(x - y)$$

$$(x + y)(x - y)$$

Therefore:

$$x^2 - y^2 = (x + y)(x - y)$$

The following two polynomials are known as the sum or difference of cubes. These are special polynomials that take the form of $x^3 + y^3$ or $x^3 - y^3$. The following formula factors the sum of cubes:

$$x^3 + y^3 = (x + y)(x^2 - xy + y^2)$$

Next is the difference of cubes, but note the change in sign. The formulas for both are similar, but the order of signs for factoring the sum or difference of cubes can be remembered by using the acronym SOAP, which stands for "same, opposite, always positive." The first sign is the same as the sign in the first expression, the second is opposite, and the third is always positive. The next formula factors the difference of cubes:

$$x^3 - y^3 = (x - y)(x^2 + xy + y^2)$$

The following two examples are expansions of cubed binomials. Similarly, these polynomials always follow a pattern:

$$x^3 + 3x^2y + 3xy^2 + y^3 = (x + y)^3$$

$$x^3 - 3x^2y + 3xy^2 - y^3 = (x - y)^3$$

These rules can be used in many combinations with one another. For example, the expression $3x^3 - 24$ has a common factor of 3, which becomes:

$$3(x^3 - 8)$$

A difference of cubes still remains which can then be factored out:

$$3(x - 2)(x^2 + 2x + 4)$$

There are no other terms to be pulled out, so this expression is completely factored.

When factoring polynomials, a good strategy is to multiply the factors to check the result. Let's try another example:

$$4x^3 + 16x^2$$

Both sides of the expression can be divided by 4, and both contain x^2, because $4x^3$ can be thought of as $4x^2(x)$, so the common term can simply be factored out:

$$4x^2(x + 4)$$

It sometimes can be necessary to rewrite the polynomial in some clever way before applying the above rules. Consider the problem of factoring $x^4 - 1$. This does not immediately look like any of the previous polynomials. However, it's possible to think of this polynomial as $x^4 - 1 = (x^2)^2 - (1^2)^2$, and now it can be treated as a difference of squares to simplify this:

$$(x^2)^2 - (1^2)^2$$

$$(x^2)^2 - x^2 1^2 + x^2 1^2 - (1^2)^2$$

$$x^2(x^2 - 1^2) + 1^2(x^2 - 1^2)$$

$$(x^2 + 1^2)(x^2 - 1^2)$$

$$(x^2 + 1)(x^2 - 1)$$

Understanding Nonlinear Relationships

A polynomial function consists of a monomial or sum of monomials arranged in descending exponential order. The graph of a polynomial function is a smooth continuous curve that extends infinitely on both ends.

The end behavior of the graph of a polynomial function can be determined by the **degree of the function** (largest exponent) and the **leading coefficient** (coefficient of the term with the largest exponent). If the degree is odd and the coefficient is positive, the graph falls to the left and rises to the right. If the degree is odd and the coefficient is negative, the graph rises to the left and falls to the right. If the degree is even and the coefficient is positive, the graph rises to the left and rises to the right. If the degree is even and the coefficient is negative, the graph falls to the left and falls to the right.

The y-intercept for any function is the point at which the graph crosses the y-axis. At this point, $x = 0$; therefore, to determine the y-intercept, substitute $x = 0$ into the function, and solve for y. Likewise, the x-intercepts, also called **zeros**, can be found by substituting $y = 0$ into the function and finding all solutions for x. For a given zero of a function, the graph can either pass through that point or simply touch that point (the graph turns at that zero). This is determined by the multiplicity of that zero. The multiplicity of a zero is the number of times its corresponding factor is multiplied to obtain the function in standard form. For example, $y = x^3 - 4x^2 - 3x + 18$ can be written in factored form as:

$$y = (x + 2)(x - 3)(x - 3) \text{ or } y = (x + 2)(x - 3)^2$$

The zeros of the function would be -2 and 3. The zero at -2 would have a multiplicity of 1, and the zero at 3 would have a multiplicity of 2. If a zero has an even multiplicity, then the graph touches the x-axis at that zero and turns around. If a zero has an odd multiplicity, then the graph crosses the x-axis at that zero.

The graph of a polynomial function can have several **turning points** (where the curve changes from rising to falling or vice versa) equal to one less than the degree of the function. For example, the function $y = 3x^5 + 2x^2 - 3x$ could have no more than four turning points.

Using Function Notation

A **function** is defined as a relationship between inputs and outputs where there is only one output value for a given input. As an example, the following function is in **function notation**:

$$f(x) = 3x - 4$$

The $f(x)$ represents the output value for an input of x. If $x = 2$, the equation becomes:

$$f(2) = 3(2) - 4 = 6 - 4 = 2$$

The input of 2 yields an output of 2, forming the ordered pair $(2, 2)$. The following set of ordered pairs corresponds to the given function: $(2, 2), (0, -4), (-2, -10)$. The set of all possible inputs of a function is its **domain**, and all possible outputs is called the **range**. By definition, each member of the domain is paired with only one member of the range.

Functions can also be defined recursively. In this form, they are not defined explicitly in terms of variables. Instead, they are defined using previously-evaluated function outputs, starting with either $f(0)$ or $f(1)$. An example of a recursively-defined function is:

$$f(1) = 2, f(n) = 2f(n-1) + 2n, n > 1$$

The domain of this function is the set of all integers.

In many cases, a function can be defined by giving an equation. For instance, $f(x) = x^2$ indicates that given a value for x, the output of f is found by squaring x.

Not all equations in x and y can be written in the form $y = f(x)$. An equation can be written in such a form if it satisfies the **vertical line test**: no vertical line meets the graph of the equation at more than a single point. In this case, y is said to be a **function of x**. If a vertical line meets the graph in two places, then this equation cannot be written in the form $y = f(x)$.

The graph of a function $f(x)$ is the graph of the equation $y = f(x)$. Thus, it is the set of all pairs (x, y) where $y = f(x)$. In other words, it is all pairs $(x, f(x))$. The x-intercepts are called the **zeros** of the function. The y-intercept is given by $f(0)$.

If, for a given function f, the only way to get $f(a) = f(b)$ is for $a = b$, then f is **one-to-one**. Often, even if a function is not one-to-one on its entire domain, it is one-to-one by considering a restricted portion of the domain.

A function $f(x) = k$ for some number k is called a **constant function**. The graph of a constant function is a horizontal line.

The function $f(x) = x$ is called the **identity function**. The graph of the identity function is the diagonal line pointing to the upper right at 45 degrees, $y = x$.

Given two functions, $f(x)$ and $g(x)$, new functions can be formed by adding, subtracting, multiplying, or dividing the functions. Any algebraic combination of the two functions can be performed, including one function being the exponent of the other. If there are expressions for f and g, then the result can be found by performing the desired operation between the expressions. So, if $f(x) = x^2$ and $g(x) = 3x$, then:

$$f \cdot g(x) = x^2 \cdot 3x = 3x^3$$

Given two functions, $f(x)$ and $g(x)$, where the domain of g contains the range of f, the two functions can be combined together in a process called **composition**. The function—"g composed of f"—is written:

$$(g \circ f)(x) = g(f(x))$$

This requires the input of x into f, then taking that result and plugging it in to the function g.

If f is one-to-one, then there is also the option to find the function $f^{-1}(x)$, called the **inverse** of f. Algebraically, the inverse function can be found by writing y in place of $f(x)$, and then solving for x. The inverse function also makes this statement true:

$$f^{-1}(f(x)) = x$$

Computing the inverse of a function f entails the following procedure:

Given $f(x) = x^2$, with a domain of $x \geq 0$

$x = y^2$ is written down o find the inverse

The square root of both sides is determined to solve for y

Normally, this would mean $\pm\sqrt{x} = y$. However, the domain of f does not include the negative numbers, so the negative option needs to be eliminated.

The result is $y = \sqrt{x}$, so $f^{-1}(x) = \sqrt{x}$, with a domain of $x \geq 0$.

A function is called **monotone** if it is either always increasing or always decreasing. For example, the functions $f(x) = 3x$ and $f(x) = -x^5$ are monotone.

An **even function** looks the same when flipped over the y-axis:

$$f(x) = f(-x)$$

The following image shows a graphic representation of an even function.

An **odd function** looks the same when flipped over the y-axis and then flipped over the x-axis:

$$f(x) = -f(-x)$$

The following image shows an example of an odd function.

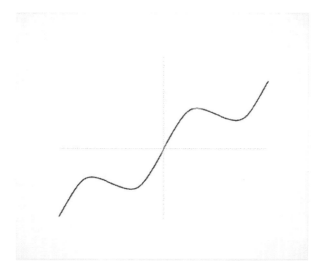

Using Structure to Isolate a Quantity of Interest

Solving equations in one variable is the process of For example, in $3x - 7 = 20$, the variable x needs to be isolated. Using opposite operations, the -7 is moved to the right side of the equation by adding seven to both sides:

$$3x - 7 + 7 = 20 + 7$$

$$3x = 27$$

Dividing by three on each side, $\frac{3x}{3} = \frac{27}{3}$, results in isolation of the variable. It is important to note that if an operation is performed on one side of the equals sign, it has to be performed on the other side to maintain equality. The solution is found to be $x = 9$. This solution can be checked for accuracy by plugging $x = 7$ in the original equation. After simplifying the equation, $20 = 20$ is found, which is a true statement.

Formulas are mathematical expressions that define the value of one quantity given the value of one or more different quantities. A formula or equation expressed in terms of one variable can be manipulated to express the relationship in terms of any other variable. The equation $y = 3x + 2$ is expressed in terms of the variable y. By manipulating the equation, it can be written as $x = \frac{y-2}{3}$, which is expressed in terms of the variable x. To manipulate an equation or formula to solve for a variable of interest, consider how the equation would be solved if all other variables were numbers. Follow the same steps for solving, leaving operations in terms of the variables instead of calculating numerical values.

The formula $P = 2l + 2w$ expresses how to calculate the perimeter of a rectangle given its length and width. To write a formula to calculate the width of a rectangle given its length and perimeter, use the previous formula relating the three variables, and solve for the variable w. If P and I were numerical values, this would be a two-step linear equation solved by subtraction and division. To solve the equation $P = 2l + 2w$ for w, first subtract 2l from both sides:

$$P - 2l = 2w$$

Then, divide both sides by 2

$$\frac{P-2l}{2} = w \text{ or } \frac{P}{2} - l = w$$

Additional Topics in Math

Solving Problems Using Volume Formulas

Geometry in three dimensions is similar to geometry in two dimensions. The main new feature is that three points now define a unique **plane** that passes through each of them. Three dimensional objects can be made by putting together two dimensional figures in different surfaces. Below, some of the possible three dimensional figures will be provided, along with formulas for their volumes.

A **rectangular prism** is a box whose sides are all rectangles meeting at 90° angles. Such a box has three dimensions: length, width, and height. If the length is *x*, the width is *y*, and the height is *z*, then the volume is given by $V = xyz$.

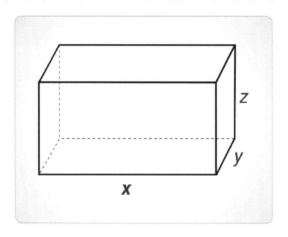

A **rectangular pyramid** is a figure with a rectangular base and four triangular sides that meet at a single vertex. If the rectangle has sides of length *x* and *y*, then the volume will be given by:

$$V = \frac{1}{3}xyh$$

A **sphere** is a set of points all of which are equidistant from some central point. It is like a circle, but in three dimensions. The volume of a sphere of radius r is given by:

$$V = \frac{4}{3}\pi r^3$$

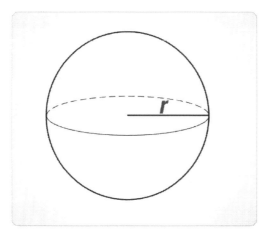

The volume of a **cylinder** is calculated by multiplying the area of the base (which is a circle) by the height of the cylinder. Doing so results in the equation:

$$V = \pi r^2 h$$

The volume of a cone is $^1/_3$ of the volume of a cylinder. Therefore, the formula for the volume of a cone is:

$$\frac{1}{3}\pi r^2 h$$

Using Trigonometric Ratios and the Pythagorean Theorem

The **Pythagorean theorem** is an important relationship between the three sides of a right triangle. It states that the square of the side opposite the right triangle, known as the **hypotenuse** (denoted as c^2), is equal to the sum of the squares of the other two sides ($a^2 + b^2$). Thus, $a^2 + b^2 = c^2$.

Both the trigonometric functions and the Pythagorean theorem can be used in problems that involve finding either a missing side or a missing angle of a right triangle. To do so, one must look to see what sides and angles are given and select the correct relationship that will help find the missing value. These relationships can also be used to solve application problems involving right triangles. Often, it's helpful to draw a figure to represent the problem to see what's missing.

To prove theorems about triangles, basic definitions involving triangles (e.g., equilateral, isosceles, etc.) need to be realized. Proven theorems concerning lines and angles can be applied to prove theorems about triangles. Common theorems to be proved include: the sum of all angles in a triangle equals 180 degrees; the sum of the lengths of two sides of a triangle is greater than the length of the third side; the base angles of an isosceles triangle are congruent; the line segment connecting the midpoint of two sides of a triangle is parallel to the third side and its length is half the length of the third side; and the medians of a triangle all meet at a single point.

An **isosceles triangle** contains at least two equal sides. Therefore, it must also contain two equal angles and, subsequently, contain two medians of the same length. An isosceles triangle can also be labelled as an **equilateral triangle** (which contains three equal sides and three equal angles) when it meets these conditions. In an equilateral triangle, the measure of each angle is always 60 degrees. Also within an equilateral triangle, the medians are of the same length. A **scalene triangle** can never be an equilateral or an isosceles triangle because it contains no equal sides and no equal angles. Also, medians in a scalene triangle can't have the same length. However, a **right triangle**, which is a triangle containing a 90-degree angle, can be a scalene triangle.

There are two types of special right triangles. The **30-60-90 right triangle** has angle measurements of 30 degrees, 60 degrees, and 90 degrees. Because of the nature of this triangle, and through the use of the Pythagorean theorem, the side lengths have a special relationship. If x is the length opposite the 30-degree angle, the length opposite the 60-degree angle is $\sqrt{3}x$, and the hypotenuse has length $2x$. The **45-45-90 right triangle** is also special as it contains two angle measurements of 45 degrees. It can be proven that, if x is the length of the two equal sides, the hypotenuse is $x\sqrt{2}$. The properties of all of these special triangles are extremely useful in determining both side lengths and angle measurements in problems where some of these quantities are given and some are not.

Trigonometric functions are also used to describe behavior in mathematics. **Trigonometry** is the relationship between the angles and sides of a triangle. **Trigonometric functions** include sine, cosine, tangent, secant, cosecant, and cotangent. The functions are defined through ratios in a right triangle. SOHCAHTOA is a common acronym used to remember these ratios, which are defined by the relationships of the sides and angles relative to the right angle. **Sine** is opposite over hypotenuse, **cosine** is adjacent over hypotenuse, and **tangent** is opposite over adjacent. These ratios are the reciprocals of secant, cosecant, and cotangent, respectively. Angles can be measured in degrees or radians. Here is a diagram of SOHCAHTOA:

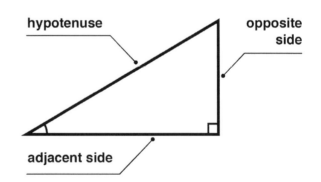

$$\text{SOH} \quad \sin\theta = \frac{\text{opposite}}{\text{hypotenuse}}$$

$$\text{CAH} \quad \cos\theta = \frac{\text{adjacent}}{\text{hypotenuse}}$$

$$\text{TOA} \quad \tan\theta = \frac{\text{opposite}}{\text{adjacent}}$$

Consider the right triangle shown in this figure:

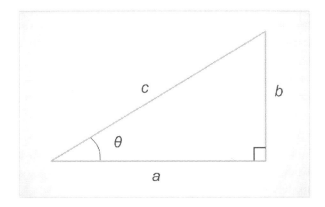

The following hold true:

- $c \sin \theta = b$

- $c \cos \theta = a$

- $\tan \theta = \frac{b}{a}$

- $b \csc \theta = c$

- $a \sec \theta = c$

- $\cot \theta = \frac{a}{b}$

A triangle that isn't a right triangle is known as an **oblique triangle**. It should be noted that even if the triangle consists of three acute angles, it is still referred to as an oblique triangle. *Oblique*, in this case, does not refer to an angle measurement. Consider the following oblique triangle:

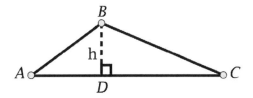

For this triangle:

$$Area = \frac{1}{2} \times base \times height = \frac{1}{2} \times AC \times BD$$

The auxiliary line drawn from the vertex B perpendicular to the opposite side AC represents the height of the triangle. This line splits the larger triangle into two smaller right triangles, which allows for the use of the trigonometric functions (specifically that $\sin A = \frac{h}{AB}$). Therefore:

$$Area = \frac{1}{2} \times AC \times AB \times \sin A$$

Typically the sides are labelled as the lowercase letter of the vertex that's opposite. Therefore, the formula can be written as $Area = \frac{1}{2}ab \sin A$. This area formula can be used to find areas of triangles when given side lengths and angle measurements, or it can be used to find side lengths or angle measurements based on a specific area and other characteristics of the triangle.

The **law of sines** and **law of cosines** are two more relationships that exist within oblique triangles. Consider a triangle with sides a, b, and c, and angles A, B, and C opposite the corresponding sides.

The law of cosines states that:

$$c^2 = a^2 + b^2 - 2ab \ \cos C$$

The law of sines states that:

$$\frac{\sin A}{a} = \frac{\sin B}{b} = \frac{\sin C}{c}$$

In addition to the area formula, these two relationships can help find unknown angle and side measurements in oblique triangles.

Complex Numbers

Complex numbers are made up of the sum of a real number and an imaginary number. **Imaginary numbers** are the result of taking the square root of -1, and $\sqrt{-1} = i$.

Some examples of complex numbers include $6 + 2i$, $5 - 7i$, and $-3 + 12i$. Adding and subtracting complex numbers is similar to collecting like terms. The real numbers are added together, and the imaginary numbers are added together. For example, if the problem asks to simplify the expression $6 + 2i - 3 + 7i$, the 6 and (-3) are combined to make 3, and the $2i$ and $7i$ combine to make $9i$. Multiplying and dividing complex numbers is similar to working with exponents. One rule to remember when multiplying is that $i * i = -1$.

For example, if a problem asks to simplify the expression $4i(3 + 7i)$, the $4i$ should be distributed throughout the 3 and the $7i$. This leaves the final expression $12i - 28$. The 28 is negative because $i * i$ results in a negative number. The last type of operation to consider with complex numbers is the conjugate. The **conjugate** of a complex number is a technique used to change the complex number into a real number. For example, the conjugate of $4 - 3i$ is $4 + 3i$. Multiplying $(4 - 3i)(4 + 3i)$ results in $16 + 12i - 12i + 9$, which has a final answer of:

$$16 + 9 = 25$$

Complex numbers may result from solving polynomial equations using the quadratic equation. Since complex numbers result from taking the square root of a negative number, the number found under the radical in the quadratic formula—called the **determinant**—tells whether or not the answer will be real or complex. If the determinant is negative, the roots are complex. Even though the coefficients of the polynomial may be real numbers, the roots are complex.

Solving polynomials by factoring is an alternative to using the quadratic formula. For example, in order to solve $x^2 - b^2 = 0$ for x, it needs to be factored. It factors into:

$$(x + b)(x - b) = 0$$

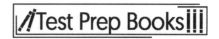

The solution set can be found by setting each factor equal to zero, resulting in $x = \pm b$. When b^2 is negative, the factors are complex numbers. For example, $x^2 + 64 = 0$ can be factored into:

$$(x + 8i)(x - 8i) = 0$$

The two roots are then found to be $x = \pm 8i$.

When dealing with polynomials and solving polynomial equations, it is important to remember the **fundamental theorem of algebra**. When given a polynomial with a degree of n, the theorem states that there will be n roots. These roots may or may not be complex. For example, the following polynomial equation of degree 2 has two complex roots: $x^2 + 1 = 0$. The factors of this polynomial are $(x + i)$ and $(x - i)$, resulting in the roots $x = i, -i$. As seen on the graph below, imaginary roots occur when the graph does not touch the x-axis.

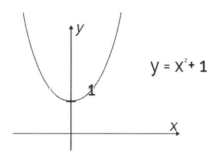

$y = x^2 + 1$

When a graphing calculator is permitted, the graph can always confirm the number and types of roots of the polynomial.

A polynomial identity is a true equation involving polynomials. For example:

$$x^2 - 5x + 6 = (x - 3)(x - 2)$$

This can be proved through multiplication by the FOIL method and factoring. This idea can be extended to involve complex numbers. For example:

$$i^2 = -1$$

$$x^3 + 9x = x(x^2 + 9) = x\left(x + \sqrt{3}i\right)\left(x - \sqrt{3}i\right)$$

This identity can also be proven through FOIL and factoring.

Converting Between Degrees and Radians

A **radian** is equal to the angle that subtends the arc with the same length as the radius of the circle. It is another unit for measuring angles, in addition to degrees. The unit circle is used to describe different radian measurements and the trigonometric ratios for special angles. The circle has a center at the origin, $(0, 0)$, and a radius of 1, which can be seen below. The points where the circle crosses an axis are labeled.

The Unit Circle

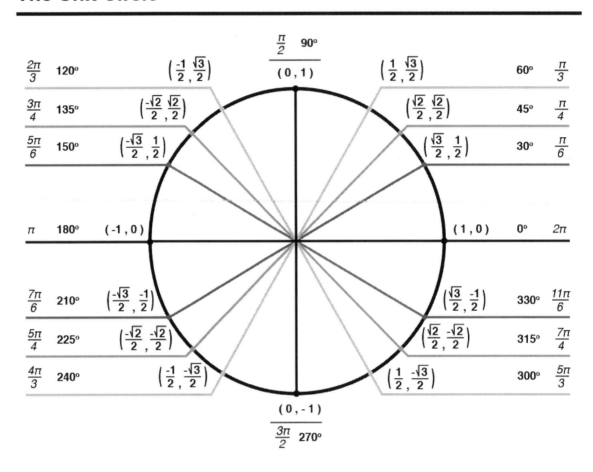

The circle begins on the right-hand side of the x-axis at 0 radians. Since the circumference of a circle is $2\pi r$ and the radius $r = 1$, the circumference is 2π. Zero and 2π are labeled as radian measurements at the point $(1, 0)$ on the graph. The radian measures around the rest of the circle are labeled also in relation to π; π is at the point $(-1, 0)$, also known as 180 degrees. Since these two measurements are equal, $\pi = 180$ degrees written as a ratio can be used to convert degrees to radians or vice versa. For example, to convert 30 degrees to radians, $30 \text{ degrees} \times \dfrac{\pi}{180 \text{ degrees}}$ can be used to obtain $\frac{1}{6}\pi$ or $\frac{\pi}{6}$. This radian measure is a point the unit circle

The coordinates labeled on the unit circle are found based on two common right triangles. The ratios formed in the coordinates can be found using these triangles. Each of these triangles can be inserted into the circle to correspond 30, 45, and 60 degrees or $\frac{\pi}{6}, \frac{\pi}{4}$, and $\frac{\pi}{3}$ radians.

By letting the hypotenuse length of these triangles equal 1, these triangles can be placed inside the unit circle. These coordinates can be used to find the trigonometric ratio for any of the radian measurements on the circle.

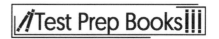

Given any (x, y) on the unit circle, $\sin(\theta) = y$, $\cos(\theta) = x$, and $\tan(\theta) = \frac{y}{x}$. The value θ is the angle that spans the arc around the unit circle. For example, finding $\sin(\frac{\pi}{4})$ means finding the y-value corresponding to the angle $\theta = \frac{\pi}{4}$. The answer is $\frac{\sqrt{2}}{2}$.

Finding $\cos(\frac{\pi}{3})$ means finding the x-value corresponding to the angle $\theta = \frac{\pi}{3}$. The answer is $\frac{1}{2}$ or 0.5. Both angles lie in the first quadrant of the unit circle. Trigonometric ratios can also be calculated for radian measures past $\frac{\pi}{2}$, or 90 degrees. Since the same special angles can be moved around the circle, the results only differ with a change in sign. This can be seen at two points labeled in the second and third quadrant.

Trigonometric functions are periodic. Both sine and cosine have period 2π. For each input angle value, the output value follows around the unit circle. Once it reaches the starting point, it continues around and around the circle. It is true that:

$$\sin(0) = \sin(2\pi) = \sin(4\pi), \text{ etc.}$$

and

$$\cos(0) = \cos(2\pi) = \cos(4\pi)$$

Tangent has period π, and its output values repeat themselves every half of the unit circle. The domain of sine and cosine are all real numbers, and the domain of tangent is all real numbers, except the points where cosine equals zero. It is also true that

$$\sin(-x) = -\sin x$$

$$\cos(-x) = \cos(x)$$

$$\tan(-x) = -\tan(x)$$

So sine and tangent are odd functions, while cosine is an even function. Sine and tangent are symmetric with respect the origin, and cosine is symmetric with respect to the y-axis.

Applying Theorems About Circles

The **radius** of a circle is the distance from the center of the circle to any point on the circle. A *chord* of a circle is a straight line formed when its endpoints are allowed to be any two points on the circle. Many angles exist within a circle. A **central angle** is formed by using two radii as its rays and the center of the circle as its vertex. An **inscribed angle** is formed by using two chords as its rays, and its vertex is a point on the circle itself. Finally, a **circumscribed angle** has a vertex that is a point outside the circle and rays that intersect with the circle. Some relationships exist between these types of angles, and, in order to define these relationships, arc measure must be understood. An **arc** of a circle is a portion of the circumference. Finding the **arc measure** is the same as finding the degree measure of the central angle that intersects the circle to form the arc. The measure of an inscribed angle is half the measure of its intercepted arc. It's also true that the measure of a circumscribed angle is equal to 180 degrees minus the measure of the central angle that forms the arc in the angle.

A **tangent line** is a line that touches a curve at a single point without going through it. A **compass** and a **straight edge** are the tools necessary to construct a tangent line from a point P outside the circle to the circle. A tangent line is constructed by drawing a line segment from the center of the circle O to the point P, and then finding its midpoint M by bisecting the line segment. By using M as the center, a compass is used to draw a circle through points O and P. N is defined as the intersection of the two circles. Finally, a line segment is drawn through P and N. This is the tangent line. Each point on a circle has only one tangent line, which is perpendicular to the radius at that point. A line similar to a tangent line is a **secant line.** Instead of intersecting the circle at one point, a secant line intersects the circle at two points. A **chord** is a smaller portion of a secant line.

Here's an example:

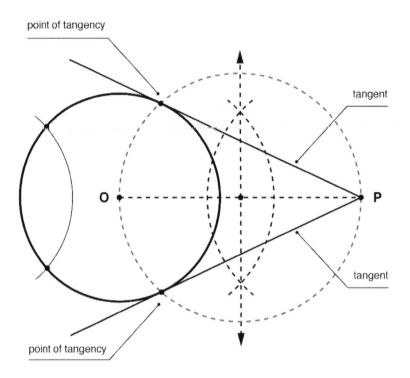

As previously mentioned, angles can be measured in radians, and 180 degrees equals π radians. Therefore, the measure of a complete circle is 2π radians. In addition to arc measure, **arc length** can also be found because the length of an arc is a portion of the circle's circumference. The following proportion is true:

$$\frac{\text{Arc measure}}{360 \text{ degrees}} = \frac{\text{arc length}}{\text{arc circumference}}$$

Arc measure is the same as the measure of the central angle, and this proportion can be rewritten as:

$$\text{arc length} = \frac{\text{central angle}}{360 \text{ degrees}} \times \text{circumference}$$

In addition, the degree measure can be replaced with radians to allow the use of both units. The arc length of a circle in radians is:

$$\text{arc length} = \text{central angle} \times \text{radius}$$

Note that arc length is a fractional part of circumference because $\frac{\text{central angle}}{360 \text{ degrees}} < 1$.

A **sector** of a circle is a portion of the circle that's enclosed by two radii and an arc. It resembles a piece of a pie, and the area of a sector can be derived using known definitions. The area of a circle can be calculated using the formula $A = \pi r^2$, where r is the radius of the circle. The area of a sector of a circle is a fraction of that calculation. For example, if the central angle θ is known in radians, the area of a sector is defined as:

$$A_s = \pi r^2 \frac{\theta}{2\pi} = \frac{\theta r^2}{2}$$

If the angle θ in degrees is known, the area of the sector is:

$$A_s = \frac{\theta \pi r^2}{360}$$

Finally, if the arc length L is known, the area of the sector can be reduced to:

$$A_s = \frac{rL}{2}$$

A chord is a line segment that contains endpoints on a circle. Given the radius of the circle and the central angle that inscribes the chord, the length of the chord can be determined.

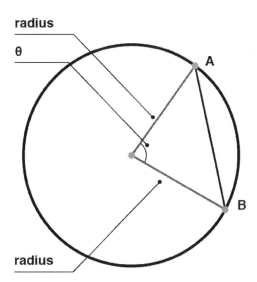

In the above figure, $\overline{AB}$ is a chord inscribed by $\angle\,\theta$. By constructing an angle bisector, the chord is also bisected at a 90 degree angle.

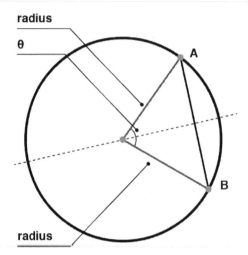

A right triangle is formed consisting of the radius as the hypotenuse, an angle equal to half the measure of θ, and a side opposite to that angle with a length half the length of the chord. Given an angle and the hypotenuse, the sin function can be used to determine the length of the side opposite the angle. Therefore, the formula $n\frac{\theta}{2}=\frac{c}{r}$, where c is the side of the triangle equal to

half of the chord, can be used. Manipulating this formula results in:

$$c = r \times \sin\frac{\theta}{2}$$

(Remember that in this formula, c represents half the length of the chord, so double this length to determine the length of the chord.)

Congruence and Similarity

Sometimes, two figures are **similar**, meaning they have the same basic shape and the same interior angles, but they have different dimensions. If the ratio of two corresponding sides is known, then that ratio, or **scale factor**, holds true for all of the dimensions of the new figure.

Here is an example of applying this principle. Suppose that Lara is 5 feet tall and is standing 30 feet from the base of a light pole, and her shadow is 6 feet long. How high is the light on the pole? To figure this, it helps to make a sketch of the situation:

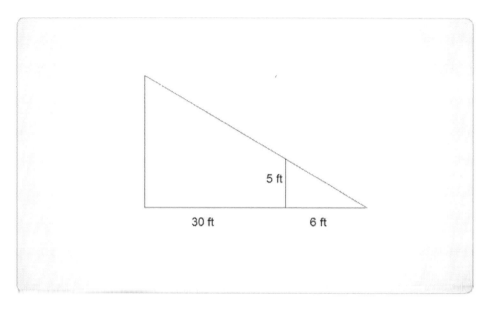

The light pole is the left side of the triangle. Lara is the 5-foot vertical line. Notice that there are two right triangles here, and that they have all the same angles as one another. Therefore, they form similar triangles. So, the ratio of proportionality between them must be determined.

The bases of these triangles are known. The small triangle, formed by Lara and her shadow, has a base of 6 feet. The large triangle formed by the light pole along with the line from the base of the pole out to the end of Lara's shadow is $30 + 6 = 36$ feet long. So, the ratio of the big triangle to the little triangle will be $\frac{36}{6} = 6$. The height of the little triangle is 5 feet. Therefore, the height of the big triangle will be $6 \times 5 = 30$ feet, meaning that the light is 30 feet up the pole.

Notice that the perimeter of a figure changes by the ratio of proportionality between two similar figures, but the area changes by the square of the ratio. This is because if the length of one side is doubled, the area is quadrupled.

As an example, suppose two rectangles are similar, but the edges of the second rectangle are three times longer than the edges of the first rectangle. The area of the first rectangle is 10 square inches. How much more area does the second rectangle have than the first?

To answer this, note that the area of the second rectangle is $3^2 = 9$ times the area of the first rectangle, which is 10 square inches. Therefore, the area of the second rectangle is going to be $9 \times 10 = 90$ square inches. This means it has $90 - 10 = 80$ square inches more area than the first rectangle.

As a second example, suppose X and Y are similar right triangles. The hypotenuse of X is 4 inches. The area of Y is $\frac{1}{4}$ the area of X. What is the hypotenuse of Y?

First, realize the area has changed by a factor of $\frac{1}{4}$. The area changes by a factor that is the square of the ratio of changes in lengths, so the ratio of the lengths is the square root of the ratio of areas. That means that the ratio of lengths must be is $\sqrt{\frac{1}{4}} = \frac{1}{2}$, and the hypotenuse of Y must be $\frac{1}{2} \times 4 = 2$ inches.

Volumes between similar solids change like the cube of the change in the lengths of their edges. Likewise, if the ratio of the volumes between similar solids is known, the ratio between their lengths is known by finding the cube root of the ratio of their volumes.

For example, suppose there are two similar rectangular pyramids X and Y. The base of X is 1 inch by 2 inches, and the volume of X is 8 inches. The volume of Y is 64 inches. What are the dimensions of the base of Y?

To answer this, first find the ratio of the volume of Y to the volume of X. This will be given by $\frac{64}{8} = 8$. Now the ratio of lengths is the cube root of the ratio of volumes, or $\sqrt[3]{8} = 2$. So, the dimensions of the base of Y must be 2 inches by 4 inches.

A **rigid motion** is a transformation that preserves distance and length. Every line segment in the resulting image is congruent to the corresponding line segment in the pre-image. **Congruence** between two figures means a series of transformations (or a rigid motion) can be defined that maps one of the figures onto the other. Basically, two figures are congruent if they have the same shape and size.

A shape is dilated, or a **dilation** occurs, when each side of the original image is multiplied by a given scale factor. If the scale factor is less than 1 and greater than 0, the dilation contracts the shape, and the resulting shape is smaller. If the scale factor equals 1, the resulting shape is the same size, and the dilation is a rigid motion. Finally, if the scale factor is greater than 1, the resulting shape is larger and the dilation expands the shape. The **center of dilation** is the point where the distance from it to any point on the new shape equals the scale factor times the distance from the center to the corresponding point in the pre-image. Dilation isn't an isometric transformation because distance isn't preserved. However, angle measure, parallel lines, and points on a line all remain unchanged.

Two figures are congruent if there is a rigid motion that can map one figure onto the other. Therefore, all pairs of sides and angles within the image and pre-image must be congruent. For example, in triangles, each pair of the three sides and three angles must be congruent. Similarly, in two four-sided figures, each pair of the four sides and four angles must be congruent.

Two figures are **similar** if there is a combination of translations, reflections, rotations, and dilations, which maps one figure onto the other. The difference between congruence and similarity is that dilation can be used in similarity. Therefore, side lengths between each shape can differ. However, angle measure must be preserved within this definition. If two polygons differ in size so that the lengths of corresponding line segments differ by the same factor, but corresponding angles have the same measurement, they are similar.

There are five theorems to show that triangles are congruent when it's unknown whether each pair of angles and sides are congruent. Each theorem is a shortcut that involves different combinations of sides and angles that must be true for the two triangles to be congruent. For example, **side-side-side (SSS)** states that if all sides are equal, the triangles are congruent. **Side-angle-side (SAS)** states that if two pairs of sides are equal and the included angles are congruent, then the triangles are congruent. Similarly, **angle-side-angle (ASA)** states that if two pairs of angles are congruent and the included side lengths are equal, the triangles are similar. **Angle-angle-side (AAS)** states that two triangles are congruent if they have two pairs of congruent angles and a pair of corresponding equal side lengths that aren't included. Finally, **hypotenuse-leg (HL)** states that if two right triangles have equal hypotenuses and an equal pair of shorter sides, then the triangles are congruent. An important item to note is that angle-angle-angle *(AAA)* is not enough information to have congruence. It's important to understand why these rules work by using rigid motions to show congruence between the triangles with the given properties. For example, three reflections are needed to show why *SAS* follows from the definition of congruence.

If two angles of one triangle are congruent with two angles of a second triangle, the triangles are similar. This is because, within any triangle, the sum of the angle measurements is 180 degrees. Therefore, if two are congruent, the third angle must also be congruent because their measurements are equal. Three congruent pairs of angles mean that the triangles are similar.

The criteria needed to prove triangles are congruent involves both angle and side congruence. Both pairs of related angles and sides need to be of the same measurement to use congruence in a proof. The criteria to prove similarity in triangles involves proportionality of side lengths. Angles must be congruent in similar triangles; however, corresponding side lengths only need to be a constant multiple of each other. Once similarity is established, it can be used in proofs as well. Relationships in geometric figures other than triangles can be proven using triangle congruence and similarity. If a similar or congruent triangle can be found within another type of geometric figure, their criteria can be used to prove a relationship about a given formula. For instance, a rectangle can be broken up into two congruent triangles.

Similarity, Triangles, and Trigonometric Ratios

Within similar triangles, corresponding sides are proportional, and angles are congruent. In addition, within similar triangles, the ratio of the side lengths is the same. This property is true even if side lengths are different. Within right triangles, trigonometric ratios can be defined for the acute angle within the triangle. The functions are defined through ratios in a right triangle. Sine of acute angle, A, is opposite over hypotenuse, cosine is adjacent over hypotenuse, and tangent is opposite over adjacent. Note that expanding or shrinking the triangle won't change the ratios. However, changing the angle measurements will alter the calculations.

Angles that add up to 90 degrees are **complementary**. Within a right triangle, two complementary angles exist because the third angle is always 90 degrees. In this scenario, the **sine** of one of the complementary angles is equal to the **cosine** of the other angle. The opposite is also true. This relationship exists because sine and cosine will be calculated as the ratios of the same side lengths.

Two lines can be parallel, perpendicular, or neither. If two lines are **parallel**, they have the same slope. This is proven using the idea of similar triangles. Consider the following diagram with two parallel lines, L1 and L2:

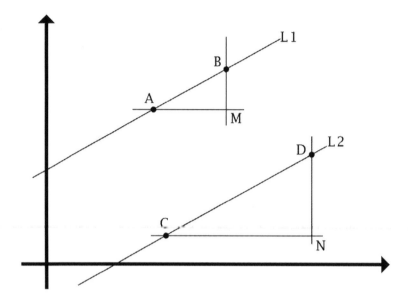

A and B are points on L1, and C and D are points on L2. Right triangles are formed with vertex M and N where lines BM and DN are parallel to the y-axis and AM and CN are parallel to the x-axis. Because all three sets of lines are parallel, the triangles are similar. Therefore, $\frac{BM}{DN} = \frac{MA}{NC}$. This shows that the rise/run is equal for lines L1 and L2. Hence, their slopes are equal.

Another similar theorem states that if there is a line parallel to one side of a triangle, and it intersects the other sides of the triangle, then the sides that are intersected are divided proportionally.

Secondly, if two lines are perpendicular, the product of their slopes equals -1. This means that their slopes are negative reciprocals of each other. Consider two perpendicular lines, l and n:

Right triangles ABC and CDE are formed so that lines BC and CE are parallel to the x-axis, and AB and DE are parallel to the y-axis. Because line BE is a straight line, angles $f + h + i = 180\ degrees$. However, angle h is a right angle, so $f + j = 90\ degrees$. By construction, $f + g = 90$, which means that $g = j$. Therefore, because angles $B = E$ and $g = j$, the triangles are similar and $\frac{AB}{BC} = \frac{CE}{DE}$. Because slope is equal to rise/run, the slope of line l is $-\frac{AB}{BC}$ and the slope of line n is $\frac{DE}{CE}$. Multiplying the slopes together gives:

$$-\frac{AB}{BC} \times \frac{DE}{CE} = -\frac{CE}{DE} \times \frac{DE}{CE} = -1$$

This proves that the product of the slopes of two perpendicular lines equals -1. Both parallel and perpendicular lines can be integral in many geometric proofs, so knowing and understanding their properties is crucial for problem-solving.

Creating Equations to Solve Problems Involving Circles

A **circle** can be defined as the set of all points that are the same distance (known as the radius, r) from a single point C (known as the center of the circle). The center has coordinates (h, k), and any point on the circle can be labelled with coordinates (x, y).

As shown below, a **right triangle** is formed with these two points:

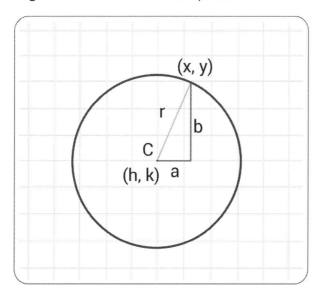

The Pythagorean theorem states that:

$$a^2 + b^2 = r^2$$

However, a can be replaced by $|x - h|$ and b can be replaced by $|y - k|$ by using the **distance formula** which is:

$$d = \sqrt{(x_2 - x_1)^2 + (y_2 - y_1)^2}$$

That substitution results in:

$$(x - h)^2 + (y - k)^2 = r^2$$

137

This is the formula for finding the equation of any circle with a center (h, k) and a radius r. Note that sometimes c is used instead of r.

Circles aren't always given in the form of the circle equation where the center and radius can be seen so easily. Oftentimes, they're given in the more general format of:

$$ax^2 + by^2 + cx + dy + e = 0$$

This can be converted to the center-radius form using the algebra technique of completing the square in both variables. First, the constant term is moved over to the other side of the equals sign, and then the x and y variable terms are grouped together. Then the equation is divided through by a and, because this is the equation of a circle, $a = b$. At this point, the x-term coefficient is divided by 2, squared, and then added to both sides of the equation. This value is grouped with the x terms. The same steps then need to be completed with the y-term coefficient. The trinomial in both x and y can now be factored into a square of a binomial, which gives both $(x - h)^2$ and $(y - k)^2$.

Practice Questions

1. Solve for x, if $x^2 - 2x - 8 = 0$.
   - a. $2 \pm \frac{\sqrt{30}}{2}$
   - b. $2 \pm 4\sqrt{2}$
   - c. 1 ± 3
   - d. $4 \pm \sqrt{2}$

2. Write the expression for three times the sum of twice a number and one minus 6.
   - a. $2x + 1 - 6$
   - b. $3x + 1 - 6$
   - c. $3(x + 1) - 6$
   - d. $3(2x + 1) - 6$

3. What is the definition of a factor of the number 36?
   - a. A number that can be divided by 36 and have no remainder
   - b. A number that 36 can be divided by and have no remainder
   - c. A prime number that is multiplied times 36
   - d. An even number that is multiplied times 36

4. What are the coordinates of the focus of the parabola $y = -9x^2$?
   - a. $(-3, 0)$
   - b. $\left(-\frac{1}{36}, 0\right)$
   - c. $(0, -3)$
   - d. $\left(0, -\frac{1}{36}\right)$

5. The total perimeter of a rectangle is 36 cm. If the length is 12 cm, what is the width?

Answer Explanations

1. C: The numbers needed are those that add to -2 and multiply to -8.

The difference between 2 and 4 is 2. Their product is 8, and -4 and 2 will work.

Therefore:

$$x^2 - 2x - 8 = (x - 4)(x + 2)$$

The latter has roots 4 and -2 or 1 ± 3.

2. D: The expression is three times the sum of twice a number and 1, which is $3(2x + 1)$. Then, 6 is subtracted from this expression.

3. B: A factor of 36 is any number that can be divided into 36 and have no remainder.

$$36 = 36 \times 1, 18 \times 2, 9 \times 4, \text{and } 6 \times 6$$

Therefore, it has 7 unique factors: 36, 18, 9, 6, 4, 2, and 1.

4. D: A parabola of the form $y = \frac{1}{4f}x^2$ has a focus $(0, f)$.

Because $y = -9x^2$, set $-9 = \frac{1}{4f}$.

Solving this equation for f results in $f = -\frac{1}{36}$. Therefore, the coordinates of the focus are $\left(0, -\frac{1}{36}\right)$.

5.

The formula for the perimeter of a rectangle is $P = 2L + 2W$, where P is the perimeter, L is the length, and W is the width. The first step is to substitute all of the data into the formula:

$$36 = 2(12) + 2W$$

Simplify by multiplying 2×12:

$$36 = 24 + 2W$$

Simplifying this further by subtracting 24 on each side, which gives:

$$36 - 24 = 24 - 24 + 2W$$

$$12 = 2W$$

Divide by 2:

$$6 = W$$

The width is 6 cm. Remember to test this answer by substituting this value into the original formula:

$$36 = 2(12) + 2(6)$$

Essay

Introduction to the SAT Essay

The essay on the SAT is now optional, although some schools do require that you write the essay for application. Test takers will have fifty minutes to write their essay. In the SAT Essay section of the exam, you will be given an essay to read and analyze. Keep in mind that you will not be asked to agree or disagree on the topic of the essay or even write about the topic of the essay. You will be asked to write about how the author built that essay and which rhetorical devices they used throughout the essay. Viewers of your essay will be looking for ways in which you explain how the author creates their argument in order to persuade an audience, and you will be expected to support your ideas with evidence from the passage.

Analyzing the Essay

In the prompt of the SAT essay, you will be asked to consider the following three characteristics of an essay: the author's evidence, reasoning, and stylistic elements. The sections below give a description of what to look for in each of these.

Evidence

It's important to evaluate the author's supporting details to be sure that the details are credible, provide evidence of the author's point, and directly support the main idea. Though shocking statistics grab readers' attention, their use could be ineffective information in the piece. Details like this are crucial to understanding the passage and evaluating how well the author presents their argument and evidence.

Fact and Opinion

It is important to distinguish between fact and opinion when reading a piece of writing. When an author presents **facts**, such as statistics or data, readers must make sure they are accurate. When authors use **opinion**, they are sharing their own thoughts and feelings about a subject.

Textual evidence within the details helps readers draw a conclusion about a passage. **Textual evidence** refers to information—facts and examples that support the main point. Textual evidence will likely come from outside sources and can be in the form of quoted or paraphrased material. In order to draw a conclusion from evidence, it's important to examine the credibility and validity of that evidence as well as how (and if) it relates to the main idea.

Credibility

Critical readers examine the facts used to support an author's argument. They check the facts against other sources to be sure those facts are correct. They also check the validity of the sources used to be sure those sources are credible, academic, and/or peer-reviewed. Consider that when an author uses another person's opinion to support their argument, even if it is an expert's opinion, it is still only an opinion and should not be taken as fact. A strong argument uses valid, measurable facts to support ideas. Even then, the reader may disagree with the argument as it may be rooted in their personal beliefs.

An authoritative argument may use the facts to sway the reader. Because of this, a writer may choose to only use the information and expert opinion that supports their viewpoint.

Reasoning

Reasoning has to do with the way the author develops ideas and connects claims and evidence. For this, it's importance to pay attention to the overall structure of the passage. If the passage is sound, the author will use effective transition words that guide the logical progression of the points. A term like *however* alerts the reader that the author is about to present information that contrasts the information presented before. Likewise, if the author were presenting cause/effect, some transition words used might be *because, since, for,* or *due to.* If these transition words are absent, then logical progression of the text is lacking and the author is using an ineffective structure to present his or her argument.

Counter-Arguments

If an author presents a differing opinion or a **counter-argument** in order to refute it, the reader should consider how and why this information is being presented. It is meant to strengthen the original argument and shouldn't be confused with the author's intended conclusion, but it should also be considered in the reader's final evaluation. On the contrary, sometimes authors will concede to an opposing argument by recognizing the validity the other side has to offer. A concession will allow readers to see both sides of the argument in an unbiased light, thereby increasing the credibility of the author.

Authors can also reflect bias if they ignore an opposing viewpoint or present their side in an unbalanced way. A strong argument considers the opposition and finds a way to refute it. Critical readers should look for an unfair or one-sided presentation of the argument and be skeptical, as a bias may be present. Even if this bias is unintentional, if it exists in the writing, the reader should be wary of the validity of the argument.

Stylistic Elements

Stylistic elements are elements an author uses in order to convey meaning to the text in unique ways. Some of these different ways are appealing to the audience's emotion, the author's word choice, and figurative language.

Appeal to Emotion

The audience is what determines the **rhetorical appeal** the author will use—ethos, pathos, or logos. **Ethos** is a rhetorical appeal to an audience's ethics and/or morals. Ethos is most often used in argumentative and informative writing modes. **Pathos** is an appeal to the audience's emotions and sympathies, and it is found in argumentative, descriptive, and narrative writing modes. **Logos** is an appeal to the audience's logic and reason and is used primarily in informative texts as well as in supporting details for argumentative pieces.

Word Choice

An author's choice of words—also referred to as **diction**—helps to convey his or her meaning in a particular way. Through diction, an author can convey a particular tone—e.g., a humorous tone, a serious tone—in order to support the thesis in a meaningful way to the reader.

Connotation refers to the implied meaning of a word or phrase. Connotations are the ideas or feelings that words or phrases invoke other than their literal meaning. An example of the use of connotation is in people's use of the word *cheap*. While the literal meaning is something of low cost, people use the word to suggest something is poor in value or to negatively describe a person as reluctant to spend money. When something or someone is described this way, the reader is more inclined to have a particular image or feeling about it or him/her. Thus, connotation can be a very effective language tool in creating emotion and swaying opinion. However, connotations are sometimes hard to pin down because varying emotions can be associated with a word. Generally, though, connotative meanings tend to be fairly consistent within a specific cultural group.

Test takers and critical readers alike should be very aware of technical language used within informational text. **Technical language** refers to terminology that is specific to a particular industry and is best understood by those specializing in that industry. This language is fairly easy to differentiate, since it will most likely be unfamiliar to readers. It's critical to be able to define technical language either by the author's written definition, through the use of an included glossary—if offered—or through context clues that help readers clarify word meaning.

Figurative Language

Literary texts also employ rhetorical devices. Figurative language like simile and metaphor is a type of rhetorical device commonly found in literature. In addition to rhetorical devices that play on the *meanings* of words, there are also rhetorical devices that use the *sounds* of words. These devices are most often found in poetry but may also be found in other types of literature and in non-fiction writing like speech texts.

Alliteration and **assonance** are both varieties of sound repetition. Other types of sound repetition include: **anaphora**, repetition that occurs at the beginning of the sentences; **epiphora**, repetition occurring at the end of phrases; **antimetabole**, repetition of words in reverse order; and **antiphrasis**, a form of denial of an assertion in a text.

Alliteration refers to the repetition of the first sound of each word. Recall Robert Burns' opening line:

> My love is like a red, red rose

This line includes two instances of alliteration: "love" and "like" (repeated *L* sound), as well as "red" and "rose" (repeated *R* sound). Next, assonance refers to the repetition of vowel sounds, and can occur anywhere within a word (not just the opening sound). Here is the opening of a poem by John Keats:

> When I have fears that I may cease to be
> Before my pen has glean'd my teeming brain

Assonance can be found in the words "fears," "cease," "be," "glean'd," and "teeming," all of which stress the long *E* sound. Both alliteration and assonance create a harmony that unifies the writer's language.

Another sound device is **onomatopoeia**, or words whose spelling mimics the sound they describe. Words such as "crash," "bang," and "sizzle" are all examples of onomatopoeia. Use of onomatopoetic language adds auditory imagery to the text.

Readers are probably most familiar with the technique of **pun**. A pun is a play on words, taking advantage of two words that have the same or similar pronunciation. Puns can be found throughout Shakespeare's plays, for instance:

> Now is the winter of our discontent
> Made glorious summer by this son of York

These lines from *Richard III* contain a play on words. Richard III refers to his brother, the newly crowned King Edward IV, as the "son of York," referencing their family heritage from the house of York. However, while drawing a comparison between the political climate and the weather (times of political trouble were the "winter," but now the new king brings "glorious summer"), Richard's use of the word "son" also implies another word with the same pronunciation, "sun"—so Edward IV is also like the sun, bringing light, warmth, and hope to England. Puns are a clever way for writers to suggest two meanings at once.

Writing the Essay

Brainstorming

One of the most important steps in writing an essay is prewriting. Before drafting an essay, it's helpful to think about the topic for a moment or two, in order to gain a more solid understanding of what the task is. Then, spending about five minutes jotting down the immediate ideas that could work for the essay is recommended. This serves as a way to get some words on the page and offer a reference for ideas when drafting. Scratch paper is provided for writers to use any prewriting techniques such as webbing, free writing, or listing. The goal is to get ideas out of the mind and onto the page.

Moving from Brainstorming to Planning

Once the ideas are on the page, it's time to turn them into a solid plan for the essay. The best ideas from the brainstorming results can then be developed into a more formal outline. An outline typically has one main point (the thesis) and at least three sub-points that support the main point.

Here's an example:

Main Idea

- Point #1
- Point #2
- Point #3

Of course, there will be details under each point, but this approach is the best for dealing with timed writing.

Time Management

It is important to manage your time effectively. It is recommended that you allocate time at the beginning of your writing to review the prompt and instructions multiple times and then outline your basic thoughts, either on paper or in your head. This initial thinking process will give you a clear plan of action before you put pen to paper and will result in a more concise and effective argument.

Similarly, it is recommended that you leave a few minutes at the end of your writing to review your piece and ensure it is coherent, includes examples that support each of your key argument points, and specifically addresses the instructions that were provided. Clarity of thought and staying focused on the topic that you were asked to write about are more important than citing every example you can think of that supports your argument in excruciating detail.

Parts of the Essay

The **introduction** has to do a few important things:

- Establish the topic of the essay in original wording (i.e., not just repeating the prompt)

- Clarify the significance/importance of the topic or purpose for writing (not too many details, a brief overview)

- Offer a **thesis statement** that identifies the writer's own viewpoint on the topic (typically one or two brief sentences as a clear, concise explanation of the main point on the topic)

Body paragraphs reflect the ideas developed in the outline. Three or four points is probably sufficient for a short essay, and they should include the following:

- A **topic sentence** that identifies the sub-point (e.g., a reason why, a way how, a cause or effect)

- A detailed explanation of the point, explaining why the writer thinks this point is valid

- Illustrative examples, such as personal examples or real-world examples, that support and validate the point (i.e., "prove" the point)

- A concluding sentence that connects the examples, reasoning, and analysis to the point being made

The **conclusion**, or final paragraph, should be brief and should reiterate the focus, clarifying why the discussion is significant or important. It is important to avoid adding specific details or new ideas to this paragraph. The purpose of the conclusion is to sum up what has been said to bring the discussion to a close.

Don't Panic!

Writing an essay can be overwhelming, and performance panic is a natural response. The outline serves as a basis for the writing and help writers stay focused. Getting stuck can also happen, and it's helpful to remember that brainstorming can be done at any time during the writing process. Following the steps of the writing process is the best defense against writer's block.

Timed essays can be particularly stressful, but assessors are trained to recognize the necessary planning and thinking for these timed efforts. Using the plan above, and sticking to it, helps with time management. Timing each part of the process helps writers stay on track. Sometimes writers try to cover too much in their essays. If time seems to be running out, writers should take the opportunity to determine whether all of the ideas in the outline are necessary. Three body paragraphs are sufficient, and more than that is probably too much to cover in a short essay.

More isn't always *better* in writing. A strong essay will be clear and concise It will avoid unnecessary or repetitive details. It is better to have a concise, five-paragraph essay that makes a clear point, than a ten-paragraph essay that doesn't. The goal is to write one to two pages of quality writing. Paragraphs should also reflect balance; if the introduction goes to the bottom of the first page, the writing may be going off-track or be repetitive. It's best to fall into the one-to-two-page range, but a complete, well-developed essay is the ultimate goal.

The Final Steps

Leaving a few minutes at the end to revise and proofread offers an opportunity for writers to polish things up. Putting oneself in the reader's shoes and focusing on what the essay actually says helps writers identify problems—it's a movement from the mindset of the writer to the mindset of the editor. The goal is to have a clean, clear copy of the essay. The following areas should be considered when proofreading:

- Sentence fragments
- Awkward sentence structure
- Run-on sentences
- Incorrect word choice
- Grammatical agreement errors
- Spelling errors
- Punctuation errors
- Capitalization errors

The Short Overview

The essay may seem challenging, but following these steps can help writers focus:

- Take one-two minutes to think about the topic.
- Generate some ideas through brainstorming (three-four minutes).
- Organize ideas into a brief outline, selecting just three-four main points to cover in the essay
- Develop essay in parts:
- Introduction paragraph, with intro to topic and main points
- Viewpoint on the subject at the end of the introduction
- Body paragraphs, based on outline
- Each paragraph: makes a main point, explains the viewpoint, uses examples to support the point
- Brief conclusion highlighting the main points and closing
- Read over the essay (last five minutes).
- Look for any obvious errors, making sure that the writing makes sense.

Essay Prompt

As you read the essay below, consider how the author uses these following things:

- Evidence to support claims, like facts or examples
- Reasoning to develop ideas and connect claims to evidence
- Stylistic elements, such as word choice or appeals to emotion, to express ideas

Dana Gioia argues in his article that poetry is dying, now little more than a limited art form confined to academic and college settings. Of course poetry remains healthy in the academic setting, but the idea of poetry being limited to this academic subculture is a stretch. New technology and social networking alone have contributed to poets and other writers' work being shared across the world. YouTube has emerged to be a major asset to poets, allowing live performances to be streamed to billions of users. Even now, poetry continues to grow and voice topics that are relevant to the culture of our time. Poetry is not in the spotlight as it may have been in earlier times, but it's still a relevant art form that continues to expand in scope and appeal.

Furthermore, Gioia's argument does not account for live performances of poetry. Not everyone has taken a poetry class or enrolled in university—but most everyone is online. The Internet is a perfect launching point to get all creative work out there. An example of this was the performance of Buddy Wakefield's *Hurling Crowbirds at Mockingbars*. Wakefield is a well-known poet who has published several collections of contemporary poetry. One of my favorite works by Wakefield is *Crowbirds*, specifically his performance at New York University in 2009. Although his reading was a campus event, views of his performance online number in the thousands. His poetry attracted people outside of the university setting.

Naturally, the poem's popularity can be attributed both to Wakefield's performance and the quality of his writing. *Crowbirds* touches on themes of core human concepts such as faith, personal loss, and growth. These are not ideas that only poets or students of literature understand, but all human beings: "You acted like I was hurling crowbirds at mockingbars / and abandoned me for not making sense. / Evidently, I don't experience things as rationally as you do" (Wakefield 15-17). Wakefield weaves together a complex description of the perplexed and hurt emotions of the speaker undergoing a separation from a romantic interest. The line "You acted like I was hurling crowbirds at mockingbars" conjures up an image of someone confused, seemingly out of their mind . . . or in the case of the speaker, passionately trying to grasp at a relationship that is fading. The speaker is looking back and finding the words that described how he wasn't making sense. This poem is particularly human and gripping in its message, but the entire effect of the poem is enhanced through the physical performance.

At its core, poetry is about addressing issues/ideas in the world. Part of this is also addressing the perspectives that are exiguously considered. Although the platform may look different, poetry continues to have a steady audience due to the emotional connection the poet shares with the audience.

Write an essay in which you explain how the author builds an argument to persuade the audience that poetry is not dying. In your essay, analyze how the author uses one or more of the features listed in the bullet points above (or features of your own choice) to strengthen the logic and persuasiveness of their argument. Focus on the most relevant features of the passage.

Practice Test #1

Reading Test

Passage #1

Questions 1-6 are based upon the following passage:

This excerpt is an adaptation of Jonathan Swift's *Gulliver's Travels into Several Remote Nations of the World.*

My gentleness and good behaviour had gained so far on the emperor and his court, and indeed upon the army and people in general, that I began to conceive hopes of getting my liberty in a short time. I took all possible methods to cultivate this favourable disposition. The natives came, by degrees, to be less apprehensive of any danger from me. I would sometimes lie down, and let five or six of them dance on my hand; and at last the boys and girls would venture to come and play at hide-and-seek in my hair. I had now made a good progress in understanding and speaking the language. The emperor had a mind one day to entertain me with several of the country shows, wherein they exceed all nations I have known, both for dexterity and magnificence. I was diverted with none so much as that of the rope-dancers, performed upon a slender white thread, extended about two feet, and twelve inches from the ground. Upon which I shall desire liberty, with the reader's patience, to enlarge a little.

This diversion is only practised by those persons who are candidates for great employments, and high favour at court. They are trained in this art from their youth, and are not always of noble birth, or liberal education. When a great office is vacant, either by death or disgrace (which often happens,) five or six of those candidates petition the emperor to entertain his majesty and the court with a dance on the rope; and whoever jumps the highest, without falling, succeeds in the office. Very often the chief ministers themselves are commanded to show their skill, and to convince the emperor that they have not lost their faculty. Flimnap, the treasurer, is allowed to cut a caper on the straight rope, at least an inch higher than any other lord in the whole empire. I have seen him do the summerset several times together, upon a trencher fixed on a rope which is no thicker than a common packthread in England. My friend Reldresal, principal secretary for private affairs, is, in my opinion, if I am not partial, the second after the treasurer; the rest of the great officers are much upon a par.

1. Which of the following statements best summarizes the central purpose of this text?
   a. Gulliver details his fondness for the archaic yet interesting practices of his captors.
   b. Gulliver conjectures about the intentions of the aristocratic sector of society.
   c. Gulliver becomes acquainted with the people and practices of his new surroundings.
   d. Gulliver's differences cause him to become penitent around new acquaintances.

2. What is the word *principal* referring to in the following text?

My friend Reldresal, principal secretary for private affairs, is, in my opinion, if I am not partial, the second after the treasurer; the rest of the great officers are much upon a par.

a. Primary or chief
b. An acolyte
c. An individual who provides nurturing
d. One in a subordinate position

3. What can the reader infer from this passage?

I would sometimes lie down, and let five or six of them dance on my hand; and at last the boys and girls would venture to come and play at hide-and-seek in my hair.

a. The children tortured Gulliver.
b. Gulliver traveled because he wanted to meet new people.
c. Gulliver is considerably larger than the children who are playing around him.
d. Gulliver has a genuine love and enthusiasm for people of all sizes.

4. What is the significance of the word *mind* in the following passage?

The emperor had a mind one day to entertain me with several of the country shows, wherein they exceed all nations I have known, both for dexterity and magnificence.

a. The ability to think
b. A collective vote
c. A definitive decision
d. A mythological question

5. Which of the following assertions does NOT support the fact that games are a commonplace event in this culture?

a. My gentleness and good behavior . . . short time.
b. They are trained in this art from their youth . . . liberal education.
c. Very often the chief ministers themselves are commanded to show their skill . . . not lost their faculty.
d. Flimnap, the treasurer, is allowed to cut a caper on the straight rope . . . higher than any other lord in the whole empire.

6. How do Flimnap and Reldresal demonstrate the community's emphasis on physical strength and leadership abilities?

a. Only children used Gulliver's hands as a playground.
b. The two men who exhibited superior abilities held prominent positions in the community.
c. Only common townspeople, not leaders, walk the straight rope.
d. No one could jump higher than Gulliver.

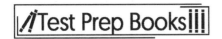

Passage #2

Questions 7-12 are based upon the following passage:

This excerpt is an adaptation of Robert Louis Stevenson's *The Strange Case of Dr. Jekyll and Mr. Hyde.*

"Did you ever come across a protégé of his—one Hyde?" He asked.

"Hyde?" repeated Lanyon. "No. Never heard of him. Since my time."

That was the amount of information that the lawyer carried back with him to the great, dark bed on which he tossed to and fro until the small hours of the morning began to grow large. It was a night of little ease to his toiling mind, toiling in mere darkness and besieged by questions.

Six o'clock struck on the bells of the church that was so conveniently near to Mr. Utterson's dwelling, and still he was digging at the problem. Hitherto it had touched him on the intellectual side alone; but now his imagination also was engaged, or rather enslaved; and as he lay and tossed in the gross darkness of the night in the curtained room, Mr. Enfield's tale went by before his mind in a scroll of lighted pictures. He would be aware of the great field of lamps in a nocturnal city; then of the figure of a man walking swiftly; then of a child running from the doctor's; and then these met, and that human Juggernaut trod the child down and passed on regardless of her screams. Or else he would see a room in a rich house, where his friend lay asleep, dreaming and smiling at his dreams; and then the door of that room would be opened, the curtains of the bed plucked apart, the sleeper recalled, and, lo! There would stand by his side a figure to whom power was given, and even at that dead hour he must rise and do its bidding. The figure in these two phrases haunted the lawyer all night; and if at anytime he dozed over, it was but to see it glide more stealthily through sleeping houses, or move the more swiftly, and still the more smoothly, even to dizziness, through wider labyrinths of lamplighted city, and at every street corner crush a child and leave her screaming. And still the figure had no face by which he might know it; even in his dreams it had no face, or one that baffled him and melted before his eyes; and thus there it was that there sprung up and grew apace in the lawyer's mind a singularly strong, almost an inordinate, curiosity to behold the features of the real Mr. Hyde. If he could but once set eyes on him, he thought the mystery would lighten and perhaps roll altogether away, as was the habit of mysterious things when well examined. He might see a reason for his friend's strange preference or bondage, and even for the startling clauses of the will. And at least it would be a face worth seeing: the face of a man who was without bowels of mercy: a face which had but to show itself to raise up, in the mind of the unimpressionable Enfield, a spirit of enduring hatred.

From that time forward, Mr. Utterson began to haunt the door in the by-street of shops. In the morning before office hours, at noon when business was plenty of time scarce, at night under the face of the full city moon, by all lights and at all hours of solitude or concourse, the lawyer was to be found on his chosen post.

"If he be Mr. Hyde," he had thought, "I should be Mr. Seek."

7. What is the purpose of the use of repetition in the following passage?

> It was a night of little ease to his toiling mind, toiling in mere darkness and besieged by questions.

    a. It serves as a demonstration of the mental state of Mr. Lanyon.
    b. It is reminiscent of the church bells that are mentioned in the story.
    c. It mimics Mr. Utterson's ambivalence.
    d. It emphasizes Mr. Utterson's anguish in failing to identify Hyde's whereabouts.

8. What is the setting of the story in this passage?
    a. In the city
    b. On the countryside
    c. In a jail
    d. In a mental health facility

9. What can one infer about the meaning of the word "Juggernaut" from the author's use of it in the passage?
    a. It is an apparition that appears at daybreak.
    b. It scares children.
    c. It is associated with space travel.
    d. Mr. Utterson finds it soothing.

10. What is the definition of the word *haunt* in the following passage?

> From that time forward, Mr. Utterson began to haunt the door in the by-street of shops. In the morning before office hours, at noon when business was plenty of time scarce, at night under the face of the full city moon, by all lights and at all hours of solitude or concourse, the lawyer was to be found on his chosen post.

    a. To levitate
    b. To constantly visit
    c. To terrorize
    d. To daunt

11. The phrase *labyrinths of lamplighted city* contains an example of what?
    a. Hyperbole
    b. Simile
    c. Juxtaposition
    d. Alliteration

12. What can one reasonably conclude from the final comment of this passage?

> "If he be Mr. Hyde," he had thought, "I should be Mr. Seek."

    a. The speaker is considering a name change.
    b. The speaker is experiencing an identity crisis.
    c. The speaker has mistakenly been looking for the wrong person.
    d. The speaker intends to continue to look for Hyde.

Passage #3

Questions 13-18 are based upon the following passage:

This excerpt is adapted from "What to the Slave is the Fourth of July?" Rochester, New York July 5, 1852

Fellow citizens—Pardon me, and allow me to ask, why am I called upon to speak here today? What have I, or those I represent, to do with your national independence? Are the great principles of political freedom and of natural justice, embodied in that Declaration of Independence, extended to us? And am I therefore called upon to bring our humble offering to the national altar, and to confess the benefits, and express devout gratitude for the blessings, resulting from your independence to us?

Would to God, both for your sakes and ours, ours that an affirmative answer could be truthfully returned to these questions! Then would my task be light, and my burden easy and delightful. For who is there so cold that a nation's sympathy could not warm him? Who so obdurate and dead to the claims of gratitude that would not thankfully acknowledge such priceless benefits? Who so stolid and selfish, that would not give his voice to swell the hallelujahs of a nation's jubilee, when the chains of servitude had been torn from his limbs? I am not that man. In a case like that, the dumb may eloquently speak, and the lame man leap as an hart.

But, such is not the state of the case. I say it with a sad sense of the disparity between us. I am not included within the pale of this glorious anniversary. Oh pity! Your high independence only reveals the immeasurable distance between us. The blessings in which you this day rejoice, I do not enjoy in common. The rich inheritance of justice, liberty, prosperity, and independence, bequeathed by your fathers, is shared by *you*, not by *me*. This Fourth of July is *yours,* not *mine.* You may rejoice, *I* must mourn. To drag a man in fetters into the grand illuminated temple of liberty, and call upon him to join you in joyous anthems, were inhuman mockery and sacrilegious irony. Do you mean, citizens, to mock me, by asking me to speak today? If so there is a parallel to your conduct. And let me warn you that it is dangerous to copy the example of a nation whose crimes, towering up to heaven, were thrown down by the breath of the Almighty, burying that nation and irrecoverable ruin! I can today take up the plaintive lament of a peeled and woe-smitten people.

By the rivers of Babylon, there we sat down. Yea! We wept when we remembered Zion. We hanged our harps upon the willows in the midst thereof. For there, they that carried us away captive, required of us a song; and they who wasted us required of us mirth, saying, "Sing us one of the songs of Zion." How can we sing the Lord's song in a strange land? If I forget thee, O Jerusalem, let my right hand forget her cunning. If I do not remember thee, let my tongue cleave to the roof of my mouth.

13. What is the tone of the first paragraph of this passage?
    a. Exasperated
    b. Inclusive
    c. Contemplative
    d. Nonchalant

14. Which word CANNOT be used synonymously with the term *obdurate* as it is conveyed in the text below?

> Who so obdurate and dead to the claims of gratitude, that would not thankfully acknowledge such priceless benefits?

    a. Steadfast
    b. Stubborn
    c. Contented
    d. Unwavering

15. What is the central purpose of this text?
    a. To demonstrate the author's extensive knowledge of the Bible
    b. To address the hypocrisy of the Fourth of July holiday
    c. To convince wealthy landowners to adopt new holiday rituals
    d. To explain why minorities often relished the notion of segregation in government institutions

16. Which statement serves as evidence of the question above?
    a. By the rivers of Babylon . . . down.
    b. Fellow citizens . . . today.
    c. I can . . . woe-smitten people.
    d. The rich Inheritance of justice . . . *not by me.*

17. The statement below features an example of which of the following literary devices?

> Oh pity! Your high independence only reveals the immeasurable distance between us.

    a. Assonance
    b. Parallelism
    c. Amplification
    d. Hyperbole

18. In the passage, the phrases "rivers of Babylon" and "songs of Zion" are considered what type of figurative language?
    a. Simile
    b. Metaphor
    c. Personification
    d. Allusion

Passage #4

Questions 19-24 are based upon the following passage:

This excerpt is an adaptation from Abraham Lincoln's Address Delivered at the Dedication of the Cemetery at Gettysburg, November 19, 1863.

> Four score and seven years ago our fathers brought forth on this continent, a new nation, conceived in liberty, and dedicated to the proposition that all men are created equal.
>
> Now we are engaged in a great civil war, testing whether that nation, or any nation so conceived and so dedicated, can long endure. We are met on a great battlefield of that

war. We have come to dedicate a portion of that field, as a final resting place for those who here gave their lives that this nation might live. It is altogether fitting and proper that we should do this.

But, in a larger sense, we cannot dedicate—we cannot consecrate that we cannot hallow—this ground. The brave men, living and dead, who struggled here, have consecrated it, far above our poor power to add or detract. The world will little note, nor long remember what we say here, but it can never forget what they did here. It is for us the living, rather, to be dedicated here to the unfinished work which they who fought here have thus far so nobly advanced. It is rather for us to be here and dedicated to the great task remaining before us—that from these honored dead we take increased devotion to that cause for which they gave the last full measure of devotion—that we here highly resolve that these dead shall not have died in vain—that this nation, under God, shall have a new birth of freedom—and that government of people, by the people, for the people, shall not perish from the earth.

19. The best description for the phrase *four score and seven years ago* is which of the following?
    a. A unit of measurement
    b. A period of time
    c. A literary movement
    d. A statement of political reform

20. What is the setting of this text?
    a. A battleship off of the coast of France
    b. A desert plain on the Sahara Desert
    c. A battlefield in North America
    d. The residence of Abraham Lincoln

21. Which war is Abraham Lincoln referring to in the following passage?
    Now we are engaged in a great civil war, testing whether that nation, or any nation so conceived and so dedicated, can long endure.

    a. World War I
    b. The War of the Spanish Succession
    c. World War II
    d. The American Civil War

22. What message is the author trying to convey through this address?
    a. The audience should perpetuate the ideals of freedom that the soldiers died fighting for.
    b. The audience should honor the dead by establishing an annual memorial service.
    c. The audience should form a militia that would overturn the current political structure.
    d. The audience should forget the lives that were lost and discredit the soldiers.

23. Which rhetorical device is being used in the following passage?

> . . . we here highly resolve that these dead shall not have died in vain—that this nation, under God, shall have a new birth of freedom—and that government of people, by the people, for the people, shall not perish from the earth.

    a. Antimetabole
    b. Antiphrasis
    c. Anaphora
    d. Epiphora

24. What is the effect of Lincoln's statement in the following passage?

> But, in a larger sense, we cannot dedicate—we cannot consecrate that we cannot hallow—this ground. The brave men, living and dead, who struggled here, have consecrated it, far above our poor power to add or detract.

    a. His comparison emphasizes the great sacrifice of the soldiers who fought in the war.
    b. His comparison serves as a reminder of the inadequacies of his audience.
    c. His comparison serves as a catalyst for guilt and shame among audience members.
    d. His comparison attempts to illuminate the great differences between soldiers and civilians.

Passage #5

Questions 25-30 are based upon the following passage:

This excerpt is adapted from Charles Dickens' speech in Birmingham in England on December 30, 1853 on behalf of the Birmingham and Midland Institute.

> My Good Friends,—When I first imparted to the committee of the projected Institute my particular wish that on one of the evenings of my readings here the main body of my audience should be composed of working men and their families, I was animated by two desires; first, by the wish to have the great pleasure of meeting you face to face at this Christmas time, and accompany you myself through one of my little Christmas books; and second, by the wish to have an opportunity of stating publicly in your presence, and in the presence of the committee, my earnest hope that the Institute will, from the beginning, recognise one great principle—strong in reason and justice—which I believe to be essential to the very life of such an Institution. It is, that the working man shall, from the first unto the last, have a share in the management of an Institution which is designed for his benefit, and which calls itself by his name.

> I have no fear here of being misunderstood—of being supposed to mean too much in this. If there ever was a time when any one class could of itself do much for its own good, and for the welfare of society—which I greatly doubt—that time is unquestionably past. It is in the fusion of different classes, without confusion; in the bringing together of employers and employed; in the creating of a better common understanding among those whose interests are identical, who depend upon each other, who are vitally essential to each other, and who never can be in unnatural antagonism without deplorable results, that one of the chief principles of a Mechanics' Institution should consist. In this world a great deal of the bitterness among us arises from an imperfect understanding of one another. Erect in Birmingham a great Educational Institution, properly educational; educational of the feelings as well as of

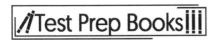

the reason; to which all orders of Birmingham men contribute; in which all orders of Birmingham men meet; wherein all orders of Birmingham men are faithfully represented—and you will erect a Temple of Concord here which will be a model edifice to the whole of England.

Contemplating as I do the existence of the Artisans' Committee, which not long ago considered the establishment of the Institute so sensibly, and supported it so heartily, I earnestly entreat the gentlemen—earnest I know in the good work, and who are now among us,—by all means to avoid the great shortcoming of similar institutions; and in asking the working man for his confidence, to set him the great example and give him theirs in return. You will judge for yourselves if I promise too much for the working man, when I say that he will stand by such an enterprise with the utmost of his patience, his perseverance, sense, and support; that I am sure he will need no charitable aid or condescending patronage; but will readily and cheerfully pay for the advantages which it confers; that he will prepare himself in individual cases where he feels that the adverse circumstances around him have rendered it necessary; in a word, that he will feel his responsibility like an honest man, and will most honestly and manfully discharge it. I now proceed to the pleasant task to which I assure you I have looked forward for a long time.

25. Which word is most closely synonymous with the word *patronage* as it appears in the following statement?

> . . . that I am sure he will need no charitable aid or condescending patronage

a. Sponsorship
b. Aberration
c. Caustic
d. Adulation

26. Which term is most closely aligned with the definition of the term *working man* as it is defined in the following passage?

> You will judge for yourselves if I promise too much for the working man, when I say that he will stand by such an enterprise with the utmost of his patience, his perseverance, sense, and support . . .

a. Plebeian
b. Viscount
c. Entrepreneur
d. Bourgeois

27. Which of the following statements most closely correlates with the definition of the term *working man* as it is defined in the previous question?
a. A working man is not someone who works for institutions or corporations, but someone who is well versed in the workings of the soul.
b. A working man is someone who is probably not involved in social activities because the physical demand for work is too high.
c. A working man is someone who works for wages among the middle class.
d. The working man has historically taken to the field, to the factory, and now to the screen.

28. Based upon the contextual evidence provided in the passage above, what is the meaning of the term *enterprise* in the third paragraph?
   a. Company
   b. Courage
   c. Game
   d. Cause

29. The speaker addresses his audience as *My Good Friends*—what kind of credibility does this salutation give to the speaker?
   a. The speaker is an employer addressing his employees, so the salutation is a way for the boss to bridge the gap between himself and his employees.
   b. The speaker's salutation is one from an entertainer to his audience and uses the friendly language to connect to his audience before a serious speech.
   c. The salutation gives the serious speech that follows a somber tone, as it is used ironically.
   d. The speech is one from a politician to the public, so the salutation is used to grab the audience's attention.

30. According to the aforementioned passage, what is the speaker's second desire for his time in front of the audience?
   a. To read a Christmas story
   b. For the working man to have a say in his institution which is designed for his benefit
   c. To have an opportunity to stand in their presence
   d. For the life of the institution to be essential to the audience as a whole

Passage #6

Questions 31-36 are based upon the following passage:

This excerpt is adapted from *Our Vanishing Wildlife,* by William T. Hornaday

> Three years ago, I think there were not many bird-lovers in the United States, who believed it possible to prevent the total extinction of both egrets from our fauna. All the known rookeries accessible to plume-hunters had been totally destroyed. Two years ago, the secret discovery of several small, hidden colonies prompted William Dutcher, President of the National Association of Audubon Societies, and Mr. T. Gilbert Pearson, Secretary, to attempt the protection of those colonies. With a fund contributed for the purpose, wardens were hired and duly commissioned. As previously stated, one of those wardens was shot dead in cold blood by a plume hunter. The task of guarding swamp rookeries from the attacks of money-hungry desperadoes to whom the accursed plumes were worth their weight in gold, is a very chancy proceeding. There is now one warden in Florida who says that "before they get my rookery they will first have to get me."

> Thus far the protective work of the Audubon Association has been successful. Now there are twenty colonies, which contain all told, about 5,000 egrets and about 120,000 herons and ibises which are guarded by the Audubon wardens. One of the most important is on Bird Island, a mile out in Orange Lake, central Florida, and it is ably defended by Oscar E. Baynard. To-day, the plume hunters who do not dare to raid the guarded rookeries are trying to study out the lines of flight of the birds, to and from

their feeding-grounds, and shoot them in transit. Their motto is—"Anything to beat the law, and get the plumes." It is there that the state of Florida should take part in the war.

The success of this campaign is attested by the fact that last year a number of egrets were seen in eastern Massachusetts—for the first time in many years. And so to-day the question is, can the wardens continue to hold the plume-hunters at bay?

31. The author's use of first person pronoun in the following text does NOT have which of the following effects?

Three years ago, I think there were not many bird-lovers in the United States, who believed it possible to prevent the total extinction of both egrets from our fauna.

a. The phrase *I think* acts as a sort of hedging, where the author's tone is less direct and/or absolute.
b. It allows the reader to more easily connect with the author.
c. It encourages the reader to empathize with the egrets.
d. It distances the reader from the text by overemphasizing the story.

32. What purpose does the quote serve at the end of the first paragraph?
a. The quote shows proof of a hunter threatening one of the wardens.
b. The quote lightens the mood by illustrating the colloquial language of the region.
c. The quote provides an example of a warden protecting one of the colonies.
d. The quote provides much needed comic relief in the form of a joke.

33. What is the meaning of the word *rookeries* in the following text?

To-day, the plume hunters who do not dare to raid the guarded rookeries are trying to study out the lines of flight of the birds, to and from their feeding-grounds, and shoot them in transit.

a. Houses in a slum area
b. A place where hunters gather to trade tools
c. A place where wardens go to trade stories
d. A colony of breeding birds

34. What is on Bird Island?
a. Hunters selling plumes
b. An important bird colony
c. Bird Island Battle between the hunters and the wardens
d. An important egret with unique plumes

35. What is the main purpose of the passage?
a. To persuade the audience to act in preservation of the bird colonies
b. To show the effect hunting egrets has had on the environment
c. To argue that the preservation of bird colonies has had a negative impact on the environment
d. To demonstrate the success of the protective work of the Audubon Association

36. Why are hunters trying to study the lines of flight of the birds?
a. To study ornithology, one must know the lines of flight that birds take.
b. To help wardens preserve the lives of the birds
c. To have a better opportunity to hunt the birds
d. To build their homes under the lines of flight because they believe it brings good luck

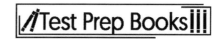

Passage #7

Objective: Read and comprehend an excerpt written from a scientific perspective.

Questions 37-42 are based upon the following passage:

This excerpt is adapted from *The Life-Story of Insects,* by Geo H. Carpenter.

Insects as a whole are preeminently creatures of the land and the air. This is shown not only by the possession of wings by a vast majority of the class, but by the mode of breathing to which reference has already been made, a system of branching air-tubes carrying atmospheric air with its combustion-supporting oxygen to all the insect's tissues. The air gains access to these tubes through a number of paired air-holes or spiracles, arranged segmentally in series.

It is of great interest to find that, nevertheless, a number of insects spend much of their time under water. This is true of not a few in the perfect winged state, as for example aquatic beetles and water-bugs ('boatmen' and 'scorpions') which have some way of protecting their spiracles when submerged, and, possessing usually the power of flight, can pass on occasion from pond or stream to upper air. But it is advisable in connection with our present subject to dwell especially on some insects that remain continually under water till they are ready to undergo their final moult and attain the winged state, which they pass entirely in the air. The preparatory instars of such insects are aquatic; the adult instar is aerial. All may-flies, dragon-flies, and caddis-flies, many beetles and two-winged flies, and a few moths thus divide their life-story between the water and the air. For the present we confine attention to the Stone-flies, the May-flies, and the Dragon-flies, three well-known orders of insects respectively called by systematists the Plecoptera, the Ephemeroptera and the Odonata.

In the case of many insects that have aquatic larvae, the latter are provided with some arrangement for enabling them to reach atmospheric air through the surface-film of the water. But the larva of a stone-fly, a dragon-fly, or a may-fly is adapted more completely than these for aquatic life; it can, by means of gills of some kind, breathe the air dissolved in water.

37. Which statement best details the central idea in this passage?
    a. It introduces certain insects that transition from water to air.
    b. It delves into entomology, especially where gills are concerned.
    c. It defines what constitutes as insects' breathing.
    d. It invites readers to have a hand in the preservation of insects.

38. Which definition most closely relates to the usage of the word *moult* in the passage?
    a. An adventure of sorts, especially underwater
    b. Mating act between two insects
    c. The act of shedding part or all of the outer shell
    d. Death of an organism that ends in a revival of life

39. What is the purpose of the first paragraph in relation to the second paragraph?
   a. The first paragraph serves as a cause, and the second paragraph serves as an effect.
   b. The first paragraph serves as a contrast to the second.
   c. The first paragraph is a description for the argument in the second paragraph.
   d. The first and second paragraphs are merely presented in a sequence.

40. What does the following sentence most nearly mean?
   The preparatory instars of such insects are aquatic; the adult instar is aerial.

   a. The volume of water is necessary to prep the insect for transition rather than the volume of the air.
   b. The abdomen of the insect is designed like a star in the water as well as the air.
   c. The stage of preparation in between molting is acted out in the water, while the last stage is in the air.
   d. These insects breathe first in the water through gills, yet continue to use the same organs to breathe in the air.

41. Which of the statements reflect information that one could reasonably infer based on the author's tone?
   a. The author's tone is persuasive and attempts to call the audience to action.
   b. The author's tone is passionate due to excitement over the subject and personal narrative.
   c. The author's tone is informative and exhibits interest in the subject of the study.
   d. The author's tone is somber, depicting some anger at the state of insect larvae.

42. Which statement best describes stoneflies, mayflies, and dragonflies?
   a. They are creatures of the land and the air.
   b. They have a way of protecting their spiracles when submerged.
   c. Their larvae can breathe the air dissolved in water through gills of some kind.
   d. The preparatory instars of these insects are aerial.

Passage #8

Questions 43-47 are based upon the following passage:

This excerpt is adapted from "The 'Hatchery' of the Sun-Fish"--- *Scientific American,* #711

I have thought that an example of the intelligence (instinct?) of a class of fish which has come under my observation during my excursions into the Adirondack region of New York State might possibly be of interest to your readers, especially as I am not aware that any one except myself has noticed it, or, at least, has given it publicity.

The female sun-fish (called, I believe, in England, the roach or bream) makes a "hatchery" for her eggs in this wise. Selecting a spot near the banks of the numerous lakes in which this region abounds, and where the water is about 4 inches deep, and still, she builds, with her tail and snout, a circular embankment 3 inches in height and 2 thick. The circle, which is as perfect a one as could be formed with mathematical instruments, is usually a foot and a half in diameter; and at one side of this circular wall an opening is left by the fish of just sufficient width to admit her body.

The mother sun-fish, having now built or provided her "hatchery," deposits her spawn within the circular inclosure, and mounts guard at the entrance until the fry are hatched out and are sufficiently large to take charge of themselves. As the embankment, moreover, is built up to the surface of the water, no enemy can very easily obtain an entrance within the inclosure from the top; while there being only one entrance, the fish is able, with comparative ease, to keep out all intruders.

I have, as I say, noticed this beautiful instinct of the sun-fish for the perpetuity of her species more particularly in the lakes of this region; but doubtless the same habit is common to these fish in other waters.

43. What is the purpose of this passage?
    a. To show the effects of fish hatcheries on the Adirondack region
    b. To persuade the audience to study Ichthyology (fish science)
    c. To depict the sequence of mating among sun-fish
    d. To enlighten the audience on the habits of sun-fish and their hatcheries

44. What does the word *wise* in this passage most closely mean?
    a. Knowledge
    b. Manner
    c. Shrewd
    d. Ignorance

45. What is the definition of the word *fry* as it appears in the following passage?
    The mother sun-fish, having now built or provided her "hatchery," deposits her spawn within the circular inclosure, and mounts guard at the entrance until the fry are hatched out and are sufficiently large to take charge of themselves.

    a. Fish at the stage of development where they are capable of feeding themselves.
    b. Fish eggs that have been fertilized.
    c. A place where larvae is kept out of danger from other predators.
    d. A dish where fish is placed in oil and fried until golden brown.

46. How is the circle that keeps the larvae of the sun-fish made?
    a. It is formed with mathematical instruments.
    b. The sun-fish builds it with her tail and snout.
    c. It is provided to her as a "hatchery" by Mother Nature.
    d. The sun-fish builds it with her larvae.

47. The author included the third paragraph in the following passage to achieve which of the following effects?
    a. To complicate the subject matter
    b. To express a bias
    c. To insert a counterargument
    d. To conclude a sequence and add a final detail

Passage #9

Questions 48-52 are based on the following passage.

In the quest to understand existence, modern philosophers must question if humans can fully comprehend the world. Classical western approaches to philosophy tend to hold that one can understand something, be it an event or object, by standing outside of the phenomena and observing it. It is then by unbiased observation that one can grasp the details of the world. This seems to hold true for many things. Scientists conduct experiments and record their findings, and thus many natural phenomena become comprehendible. However, several of these observations were possible because humans used tools in order to make these discoveries.

This may seem like an extraneous matter. After all, people invented things like microscopes and telescopes in order to enhance their capacity to view cells or the movement of stars. While humans are still capable of seeing things, the question remains if human beings have the capacity to fully observe and see the world in order to understand it. It would not be an impossible stretch to argue that what humans see through a microscope is not the exact thing itself, but a human interpretation of it.

This would seem to be the case in the "Business of the Holes" experiment conducted by Richard Feynman. To study the way electrons behave, Feynman set up a barrier with two holes and a plate. The plate was there to indicate how many times the electrons would pass through the hole(s). Rather than casually observe the electrons acting under normal circumstances, Feynman discovered that electrons behave in two totally different ways depending on whether or not they are observed. The electrons that were observed had passed through either one of the holes or were caught on the plate as particles. However, electrons that weren't observed acted as waves instead of particles and passed through both holes. This indicated that electrons have a dual nature. Electrons seen by the human eye act like particles, while unseen electrons act like waves of energy.

This dual nature of the electrons presents a conundrum. While humans now have a better understanding of electrons, the fact remains that people cannot entirely perceive how electrons behave without the use of instruments. We can only observe one of the mentioned behaviors, which only provides a partial understanding of the entire function of electrons. Therefore, we're forced to ask ourselves whether the world we observe is objective or if it is subjectively perceived by humans. Or, an alternative question: can man understand the world only through machines that will allow them to observe natural phenomena?

Both questions humble man's capacity to grasp the world. However, those ideas don't take into account that many phenomena have been proven by human beings without the use of machines, such as the discovery of gravity. Like all philosophical questions, whether man's reason and observation alone can understand the universe can be approached from many angles.

48. The word *extraneous* in paragraph two can be best interpreted as referring to which one of the following?
    a. Indispensable
    b. Bewildering
    c. Superfluous
    d. Exuberant

49. What is the author's motivation for writing the passage?
    a. Bring to light an alternative view on human perception by examining the role of technology in human understanding.
    b. Educate the reader on the latest astroparticle physics discovery and offer terms that may be unfamiliar to the reader.
    c. Argue that humans are totally blind to the realities of the world by presenting an experiment that proves that electrons are not what they seem on the surface.
    d. Reflect on opposing views of human understanding.

50. Which of the following most closely resembles the way in which paragraph four is structured?
    a. It offers one solution, questions the solution, and then ends with an alternative solution.
    b. It presents an inquiry, explains the detail of that inquiry, and then offers a solution.
    c. It presents a problem, explains the details of that problem, and then ends with more inquiry.
    d. It gives a definition, offers an explanation, and then ends with an inquiry.

51. For the classical approach to understanding to hold true, which of the following must be required?
    a. A telescope.
    b. The person observing must prove their theory beyond a doubt.
    c. Multiple witnesses present.
    d. The person observing must be unbiased.

52. Which best describes how the electrons in the experiment behaved like waves?
    a. The electrons moved up and down like actual waves.
    b. The electrons passed through both holes and then onto the plate.
    c. The electrons converted to photons upon touching the plate.
    d. Electrons were seen passing through one hole or the other.

Writing and Language Test

Careers

Focus: Students must make revising and editing decisions on the context of a passage on a career-related topic.

Aircraft Engineers

The knowledge of an aircraft engineer is acquired through years of education, and special licenses are required. Ideally, an individual will begin his or her preparation for the profession in high school by taking chemistry, physics, trigonometry, and calculus. Such curricula will aid in one's pursuit of a bachelor's degree in aircraft engineering, which requires several physical and life sciences, mathematics, and design courses.

(2) <u>Some of universities provide internship or apprentice opportunities</u> for the students enrolled in aircraft engineer programs. A bachelor's in aircraft engineering is commonly accompanied by a master's degree in advanced engineering or business administration. Such advanced degrees enable an individual to position himself or herself for executive, faculty, and/or research opportunities. (3) <u>These advanced offices oftentimes require a Professional Engineering (PE) license which can be obtained through additional college courses, professional experience, and acceptable scores on the Fundamentals of Engineering (FE) and Professional Engineering (PE) standardized assessments.</u>

Once the job begins, this line of work requires critical thinking, business skills, problem solving, and creativity. This level of (5) <u>expertise</u> (6) <u>allows</u> aircraft engineers to apply mathematical equations and scientific processes to aeronautical and aerospace issues or inventions. (8) <u>For example, aircraft engineers may test, design, and construct flying vessels such as airplanes, space shuttles, and missile weapons.</u> As a result, aircraft engineers are compensated with generous salaries. In fact, in May 2014, the lowest 10 percent of all American aircraft engineers earned less than $60,110 while the highest paid ten-percent of all American aircraft engineers earned $155,240. (9) <u>In May 2015, the United States Bureau of Labor Statistics (BLS) reported that the median annual salary of aircraft engineers was $107,830.</u> (10) <u>Conversely,</u> (11) <u>employment opportunities for aircraft engineers are projected to decrease by 2 percent by 2024.</u> This decrease may be the result of a decline in the manufacturing industry. Nevertheless, aircraft engineers who know how to utilize modeling and simulation programs, fluid dynamic software, and robotic engineering tools are projected to remain the most employable.

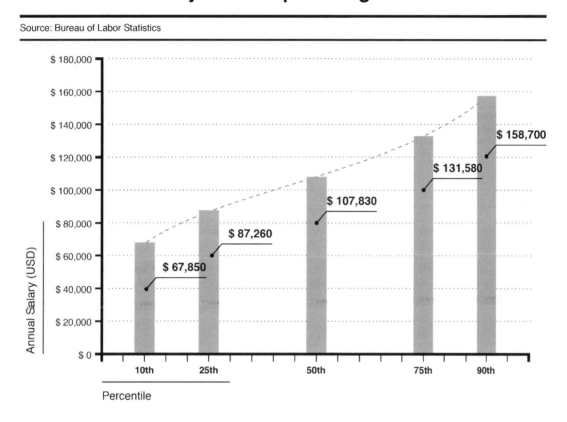

2015 Annual Salary of Aerospace Engineers

Source: Bureau of Labor Statistics

1. What type of text is utilized in the passage?
   a. Argumentative
   b. Narrative
   c. Biographical
   d. Informative

2. Which of the following would be the best choice for this sentence (reproduced below)?

   (2) <u>Some of universities provide internship or apprentice opportunities</u> for the students enrolled in aircraft engineer programs.

   a. NO CHANGE
   b. Some of universities provided internship or apprentice opportunities
   c. Some of universities provide internship or apprenticeship opportunities
   d. Some universities provide internship or apprenticeship opportunities

3. Which of the following would be the best choice for this sentence (reproduced below)?

(3) <u>These advanced offices oftentimes require a Professional Engineering (PE) license which can be obtained through additional college courses, professional experience, and acceptable scores on the Fundamentals of Engineering (FE) and Professional Engineering (PE) standardized assessments.</u>

a. NO CHANGE
b. These advanced positions oftentimes require acceptable scores on the Fundamentals of Engineering (FE) and Professional Engineering (PE) standardized assessments in order to achieve a Professional Engineering (PE) license. Additional college courses and professional experience help.
c. These advanced offices oftentimes require acceptable scores on the Fundamentals of Engineering (FE) and Professional Engineering (PE) standardized assessments to gain the Professional Engineering (PE) license which can be obtained through additional college courses, professional experience.
d. These advanced positions oftentimes require a Professional Engineering (PE) license which is obtained by acceptable scores on the Fundamentals of Engineering (FE) and Professional Engineering (PE) standardized assessments. Further education and professional experience can help prepare for the assessments.

4. "The knowledge of an aircraft engineer is acquired through years of education." Which statement serves to support this claim?
    a. Aircraft engineers are compensated with generous salaries.
    b. Such advanced degrees enable an individual to position himself or herself for executive, faculty, or research opportunities.
    c. Ideally, an individual will begin his or her preparation for the profession in high school by taking chemistry, physics, trigonometry, and calculus.
    d. Aircraft engineers who know how to utilize modeling and simulation programs, fluid dynamic software, and robotic engineering tools will be the most employable.

5. What is the meaning of "expertise" in the marked sentence?
    a. Care
    b. Skill
    c. Work
    d. Composition

6. Which of the following would be the best choice for this sentence (reproduced below)?

This level of expertise (6) <u>allows</u> aircraft engineers to apply mathematical equation and scientific processes to aeronautical and aerospace issues or inventions.

a. NO CHANGE
b. Inhibits
c. Requires
d. Should

7. In the third paragraph, which of the following claims is supported?
   a. This line of work requires critical thinking, business skills, problem solving, and creativity.
   b. Aircraft engineers are compensated with generous salaries.
   c. The knowledge of an aircraft engineer is acquired through years of education.
   d. Those who work hard are rewarded accordingly.

8. Which of the following would be the best choice for this sentence (reproduced below)?

   (8) <u>For example,</u> aircraft engineers may test, design, and construct flying vessels such as airplanes, space shuttles, and missile weapons.

   a. NO CHANGE
   b. Therefore,
   c. However,
   d. Furthermore,

9. Which of the following would be the best choice for this sentence (reproduced below)?

   (9) <u>In May 2015, the United States Bureau of Labor Statistics (BLS) reported that the median annual salary of aircraft engineers was $107,830.</u>

   a. NO CHANGE
   b. May of 2015, the United States Bureau of Labor Statistics (BLS) reported that the median annual salary of aircraft engineers was $107,830.
   c. In May of 2015 the United States Bureau of Labor Statistics (BLS) reported that the median annual salary of aircraft engineers was $107,830.
   d. In May, 2015, the United States Bureau of Labor Statistics (BLS) reported that the median annual salary of aircraft engineers was $107,830.

10. Which of the following would be the best choice for this sentence (reproduced below)?

    (10) <u>Conversely,</u> employment opportunities for aircraft engineers are projected to decrease by 2 percent by 2024.

    a. NO CHANGE
    b. Similarly,
    c. In other words,
    d. Accordingly,

11. Which of the following would be the best choice for this sentence (reproduced below)?

    Conversely, (11) <u>employment opportunities for aircraft engineers are projected to decrease by 2 percent by 2024.</u>

    a. NO CHANGE
    b. Employment opportunities for aircraft engineers will be projected to decrease by 2 percent by 2024.
    c. Employment opportunities for aircraft engineers is projected to decrease by 2 percent by 2024.
    d. Employment opportunities for aircraft engineers was projected to decrease by 2 percent by 2024.

History/ Social Studies

Focus: Students must make revising and editing decisions on the context of a passage on a history/social studies topic.

Attacks of September 11th

On September 11th 2001, a group of terrorists hijacked four American airplanes. The terrorists crashed the planes into the World Trade Center in New York City, the Pentagon in Washington D.C., and a field in Pennsylvania. Nearly 3,000 people died during the attacks, which propelled the United States into a "War on Terror".

About the Terrorists

Terrorists commonly use fear and violence to achieve political goals. The nineteen terrorists who orchestrated and implemented the attacks of September 11th were militants associated with al-Qaeda, an Islamic extremist group founded by Osama bin Laden, Abdullah Azzam, and others in the late 1980s. (13) Bin Laden orchestrated the attacks as a response to what he felt was American injustice against Islam and hatred towards Muslims. In his words, "Terrorism against America deserves to be praised."

Islam is the religion of Muslims, who live mainly in South and Southwest Asia and Sub-Saharan Africa. The majority of Muslims practice Islam peacefully. However, fractures in Islam have led to the growth of Islamic extremists who strictly oppose Western influences. They seek to institute stringent Islamic law and destroy those who (15) violate Islamic code.

In November 2002, bin Laden provided the explicit motives for the 9/11 terror attacks. According to this list, America's support of Israel, military presence in Saudi Arabia, and other anti-Muslim actions were the causes.

The Timeline of the Attacks

The morning of September 11 began like any other for most Americans. Then, at 8:45 a.m., a Boeing 767 plane crashed into the north tower of the World Trade Center in New York City. Hundreds were instantly killed. Others were trapped on higher floors. The (17) crash was initially thought to be a freak accident. When a second plane flew directly into the south tower eighteen minutes later, it was determined that America was under attack.

At 9:45 a.m., a third plane slammed into the Pentagon, America's military headquarters in Washington D.C. The jet fuel of this plane caused a major fire and partial building collapse that resulted in nearly 200 deaths. By 10:00 a.m., the south tower of the World Trade Center collapsed. Thirty minutes later, the north tower followed suit.

While this was happening, a fourth plane that departed from New Jersey, United Flight 93, was hijacked. The passengers learned of the attacks that occurred in New York and Washington D.C. and realized that they faced the same fate as the other planes that crashed. The passengers were determined to overpower the terrorists in an effort to prevent the deaths of additional innocent American citizens. Although the passengers were successful in (18) <u>diverging</u> the plane, it crashed in a western Pennsylvania field and killed everyone on board. The plane's final target remains uncertain, but many believe that United Flight 93 was heading for the White House.

Heroes and Rescuers

Close to 3,000 people died in the World Trade Center attacks. This figure includes 343 New York City firefighters and paramedics, 23 New York City police officers, and 37 Port Authority officers. Nevertheless, thousands of men and women in service worked valiantly to evacuate the buildings, save trapped workers, extinguish infernos, uncover victims trapped in fallen rubble, and tend to nearly 10,000 injured individuals.

About 300 rescue dogs played a major role in the after-attack salvages. Working twelve-hour shifts, the dogs scoured the rubble and alerted paramedics when they found signs of life. While doing so, the dogs served as a source of comfort and therapy for the rescue teams.

Initial Impacts on America

The attacks of September 11, 2001 resulted in the immediate suspension of all air travel. No flights could take off from or land on American soil. American airports and airspace closed to all national and international flights. Therefore, over five hundred flights had to turn back or be redirected to other countries. Canada alone received 226 flights and thousands of stranded passengers. Needless to say, as canceled flights are rescheduled, air travel became backed up and chaotic for quite some time.

At the time of the attacks, George W. Bush was the president of the United States. President Bush announced that "We will make no distinction between the terrorists who committed these acts and those who harbor them." The rate of hate crimes against American Muslims spiked, despite President Bush's call for the country to treat them with respect.

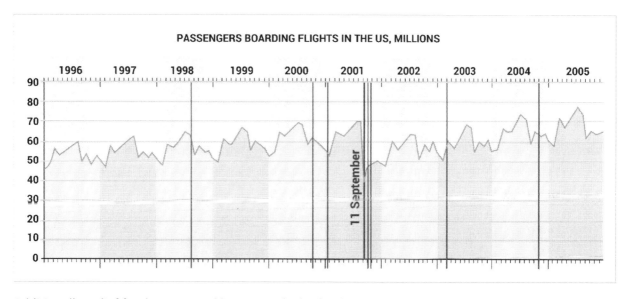

PASSENGERS BOARDING FLIGHTS IN THE US, MILLIONS

Additionally, relief funds were quickly arranged. The funds were used to support families of the victims, orphaned children, and those with major injuries. In this way, the

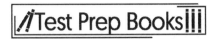

tragic event brought the citizens together through acts of service towards those directly impacted by the attack.

Long-term Effects of the Attacks

Over the past fifteen years, the attacks of September 11th have transformed the United States' government, travel safety protocols, and international relations. Anti-terrorism legislation became a priority for many countries as law enforcement and intelligence agencies teamed up to find and defeat alleged terrorists.

Present George W. Bush announced a War on Terror. He (21) desired to bring bin Laden and al-Qaeda to justice and prevent future terrorist networks from gaining strength. The War in Afghanistan began in October of 2001 when the United States and British forces bombed al-Qaeda camps. (22) The Taliban, a group of fundamental Muslims who protected Osama bin Laden, was overthrown on December 9, 2001. However, the war continued in order to defeat insurgency campaigns in neighboring countries. Ten years later, the United State Navy SEALS killed Osama bin Laden in Pakistan. During 2014, the United States declared the end of its involvement in the War on Terror in Afghanistan.

Museums and memorials have since been erected to honor and remember the thousands of people who died during the September 11th attacks, including the brave rescue workers who gave their lives in the effort to help others.

12. How does the structure of the text help readers better understand the topic?
a. By stating that anti-terrorism legislation was a priority for many countries, the reader can determine which laws were made and how they changed the life in the country.
b. By placing the events in the order that they occurred, readers are better able to understand how the day unfolded.
c. By using descriptive language, the readers are able to develop detailed images of the events that occurred during September 11, 2001.
d. None of the above

13. Which of the following would be the best choice for this sentence (reproduced below)?

(13) Bin Laden orchestrated the attacks as a response to what he felt was American injustice against Islam and hatred towards Muslims.

a. NO CHANGE
b. Bin Laden orchestrated the attacks as a response to what he felt was American injustice against Islam, and hatred towards Muslims.
c. Bin Laden orchestrated the attacks, as a response to what he felt was American injustice against Islam and hatred towards Muslims.
d. Bin Laden orchestrated the attacks as responding to what he felt was American injustice against Islam and hatred towards Muslims.

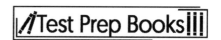

14. How does the author express that most Muslims are peaceful people?
    a. By describing the life of a Muslim after the attacks.
    b. By including an anecdote about a Muslim friend.
    c. By reciting details from religious texts.
    d. By explicitly stating that fact.

15. What word could be used in exchange for "violate"?
    a. Respect
    b. Defile
    c. Deny
    d. Obey

16. What technique does the author use to highlight the impact of United Flight 93?
    a. An image of the crash
    b. An allusion to illustrate what may have occurred had the passengers not taken action
    c. An anecdote about a specific passenger
    d. A point of view consideration, where the author forces the reader to think about how he or she would have responded to such a situation

17. Which of the following would NOT be an appropriate replacement for the underlined portion of the sentence (reproduced below)?

    The (17) crash was initially thought to be a freak accident.

    a. First crash was thought to be
    b. Initial crash was thought to be
    c. Thought was that the initial crash
    d. Initial thought was that the crash was

18. Which of the following would be the best choice for this sentence (reproduced below)?

    Although the passengers were successful in (18) diverging the plane, it crashed in a western Pennsylvania field and killed everyone on board.

    a. NO CHANGE
    b. Diverting
    c. Converging
    d. Distracting

19. What statement is best supported by the graph included in this passage?
    a. As canceled flights were rescheduled, air travel became backed up and chaotic for quite some time.
    b. Over five hundred flights had to turn back or be redirected to other countries.
    c. Canada alone received 226 flights and thousands of stranded passengers.
    d. In the first few months following the attacks, there was a significant decrease in passengers boarding flights.

20. What is the purpose of the last paragraph?
    a. It shows that beautiful art can be used to remember a past event.
    b. It demonstrates that Americans will always remember the 9/11 attacks and the lives that were lost.
    c. It explains how America fought back after the attacks.
    d. It provides the author with an opportunity to explain how the location of the towers is used today.

21. Which of the following would be the best choice for this sentence (reproduced below)?

    He (21) <u>desired</u> to bring bin Laden and al-Qaeda to justice and prevent future terrorist networks from gaining strength.

    a. NO CHANGE
    b. Perceived
    c. Intended
    d. Assimilated

22. Which of the following would be the best choice for this sentence (reproduced below)?

    (22) <u>The Taliban, a group of fundamental Muslims who protected Osama bin Laden, was overthrown on December 9, 2001. However, the war continued in order to defeat insurgency campaigns in neighboring countries.</u>

    a. NO CHANGE
    b. The Taliban was overthrown on December 9, 2001. They were a group of fundamental Muslims who protected Osama bin Laden. However, the war continued in order to defeat insurgency campaigns in neighboring countries.
    c. The Taliban, a group of fundamental Muslims who protected Osama bin Laden, on December 9, 2001 was overthrown. However, the war continued in order to defeat insurgency campaigns in neighboring countries.
    d. Osama bin Laden's fundamental Muslims who protected him were called the Taliban and overthrown on December 9, 2001. Yet the war continued in order to defeat the insurgency campaigns in neighboring countries.

Humanities

Focus: Students must make revising and editing decisions on the context of a passage on a humanities-related topic.

Fred Hampton

Fred Hampton desired to see lasting social change for African American people through nonviolent means and community recognition. <u>(23) As a result, he became an African American activist</u> during the American Civil Rights Movement and led the Chicago chapter of the Black Panther Party.

Hampton's Education

Hampton was born and raised in Maywood of Chicago, Illinois in 1948. (24) Gifted academically and a natural athlete, he became a stellar baseball player in high school. After graduating from Proviso East High School in 1966, he later went on to study law at Triton Junior College.

While studying at Triton, Hampton joined and became a leader of the National Association for the Advancement of Colored People (NAACP). (25) As a result of his leadership, the NAACP gained more than 500 members. Hampton worked relentlessly to acquire recreational facilities in the neighborhood and improve the educational resources provided to the impoverished black community of Maywood.

The Black Panthers

The Black Panther Party (BPP) was another activist group that formed around the same time as the NAACP. Hampton was quickly attracted to the Black Panther's approach to the fight for equal rights for African Americans. (26) Hampton eventually joined the chapter and relocated to downtown Chicago to be closer to its headquarters.

His (27) charismatic personality, organizational abilities, sheer determination, and rhetorical skills enabled him to quickly rise through the chapter's ranks. (28) Hampton soon became the leader of the Chicago chapter of the BPP where he organized rallies, taught political education classes, and established a free medical clinic. He also took part in the community police supervision project and played an instrumental role in the BPP breakfast program for impoverished African American children.

Hampton's greatest achievement as the leader of the BPP may be his fight against street gang violence in Chicago. In 1969, Hampton held a press conference where he made the gangs agree to a nonaggression pact known as the Rainbow Coalition. As a result of the pact, a multiracial alliance between blacks, Puerto Ricans, and poor youth was developed.

Assassination

As the Black Panther Party's popularity and influence grew, the Federal Bureau of Investigation (FBI) placed the group under constant surveillance. In an attempt to (30) neutralize the party, the FBI launched several harassment campaigns against the BPP, raided its headquarters in Chicago three times, and arrested over one hundred of the group's members. Hampton was shot during such a raid that occurred on the morning of December 4th, 1969.

In 1976, seven years after the event, it was revealed that William O'Neal, Hampton's trusted bodyguard, was an undercover FBI agent. (31) O'Neal provided the FBI with detailed floor plans of the BPP's headquarters, identifying the exact location of Hampton's bed. It was because of these floor plans that the police were able to target and kill Hampton.

The assassination of Hampton fueled outrage amongst the African American community. It was not until years after the assassination that the police admitted wrongdoing. The Chicago City Council now (32) commemorates December 4th as Fred Hampton Day.

23. Which of the following would be the best choice for this sentence (reproduced below)?

(23) As a result, he became an African American activist during the American Civil Rights Movement and led the Chicago chapter of the Black Panther Party.

a. NO CHANGE
b. As a result he became an African American activist
c. As a result: he became an African American activist
d. As a result of, he became an African American activist

24. What word could be used in place of the underlined description?

24) Gifted academically and a natural athlete, he became a stellar baseball player in high school.

a. Vacuous
b. Energetic
c. Intelligent
d. Athletic

25. Which of the following statements, if true, would further validate the selected sentence?
a. Several of these new members went on to earn scholarships.
b. With this increase in numbers, Hampton was awarded a medal for his contribution to the NAACP.
c. This increase in membership was unprecedented in the NAACP's history.
d. The NAACP has been growing steadily every year.

26. How else could this sentence be re-structured while maintaining the context of the fourth paragraph?
a. NO CHANGE
b. Eventually, Hampton joined the chapter and relocated to downtown Chicago to be closer to its headquarters.
c. Nevertheless, Hampton joined the chapter and relocated to downtown Chicago to be closer to its headquarters.
d. Hampton then joined the chapter and relocated to downtown Chicago to be closer to its headquarters

27. What word is synonymous with the underlined description?
a. Egotistical
b. Obnoxious
c. Chauvinistic
d. Charming

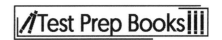

28. Which of the following would be the best choice for this sentence (reproduced below)?

(28) <u>Hampton soon became the leader of the Chicago chapter of the BPP where he organized rallies, taught political education classes, and established a free medical clinic.</u>

a. NO CHANGE
b. As the leader of the BPP, Hampton: organized rallies, taught political education classes, and established a free medical clinic.
c. As the leader of the BPP, Hampton; organized rallies, taught political education classes, and established a free medical clinic.
d. As the leader of the BPP, Hampton—organized rallies, taught political education classes, and established a medical free clinic.

29. The author develops the idea that Frank Hampton should not have been killed at the hands of the police. Which could best be used to support that claim?
a. The manner in which the police raided the BPP headquarters.
b. The eventual admission from the police that they were wrong in killing Hampton.
c. The description of previous police raids that resulted in the arrest of hundreds BPP members.
d. All of the above.

30. Which of the following would be the best choice for this sentence (reproduced below)?

In an attempt to (30) <u>neutralize</u> the party, the FBI launched several harassment campaigns against the BPP, raided its headquarters in Chicago three times, and arrested over one hundred of the group's members.

a. NO CHANGE
b. Accommodate
c. Assuage
d. Praise

31. Which of the following would be the best choice for this sentence (reproduced below)?

(31) <u>O'Neal provided the FBI with detailed floor plans of the BPP's headquarters, identifying the exact location of Hampton's bed.</u>

a. NO CHANGE
b. The FBI provided O'Neal with detailed floor plans of the BPP's headquarters, which identified the exact location of Hampton's bed.
c. O'Neal provided the FBI with detailed floor plans and Hampton's bed.
d. O'Neal identified the exact location of Hampton's bed that provided the FBI with detailed floor plans of the BPP's headquarters.

32. What word could be used in place of the underlined word?
a. Disregards
b. Memorializes
c. Communicates
d. Denies

33. How would the author likely describe the FBI during the events of the passage?
    a. Corrupt
    b. Confused
    c. Well-intended
    d. Prejudiced

Science

Focus: Students must make revising and editing decisions on the context of a passage on a scientific topic.

Here Comes the Flood!

A flood occurs when an area of land that is normally dry becomes submerged with water. Floods have affected Earth since the beginning of time and are caused by many different factors. (36) Flooding can occur slowly or within seconds and can submerge small regions or extend over vast areas of land. Their impact on society and the environment can be harmful or helpful.

What Causes Flooding?

Floods may be caused by natural phenomenon, induced by the activities of humans and other animals, or the failure of an infrastructure. Areas located near bodies of water are prone to flooding as are low-lying regions.

Global warming is the result of air pollution that prevents the sun's radiation from being emitted back into space. Instead, the radiation is trapped in Earth and results in global warming. The warming of the Earth has resulted in climate changes. As a result, floods have been occurring with increasing regularity. Some claim that the increased temperatures on Earth may cause the icebergs to melt. They fear that the melting of icebergs will cause the (37) oceans levels to rise and flood coastal regions.

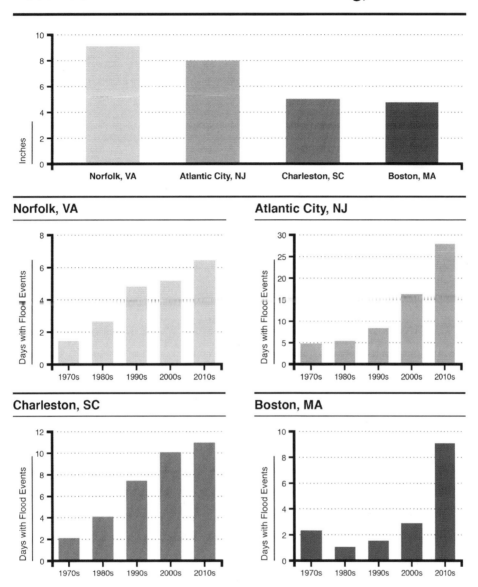

Most commonly, flooding is caused by excessive rain. The ground is not able to absorb all the water produced by a sudden heavy rainfall or rainfall that occurs over a prolonged period of time. Such rainfall may cause the water in rivers and other bodies of water to overflow. The excess water can cause dams to break. Such events can cause flooding of the surrounding riverbanks or coastal regions.

Flash flooding can occur without warning and without rainfall. Flash floods may be caused by a river being blocked by a glacier, avalanche, landslide, logjam, a beaver's obstruction, construction, or dam. Water builds behind such a blockage. Eventually, the mass and force of the built-up water become so extreme that it causes the obstruction to break. Thus, enormous amounts of water rush out towards the surrounding areas.

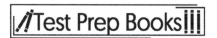

Areal or urban flooding occurs because the land has become hardened. The hardening of land may result from urbanization or drought. Either way, the hardened land prevents water from seeping into the ground. Instead, the water resides on top of the land.

Finally, flooding may result after severe hurricanes, tsunamis, or tropical cyclones. Local defenses and infrastructures are no matches for the tidal surges and waves caused by these natural phenomena. Such events are bound to result in the flooding of nearby coastal regions or estuaries.

A Floods After-Effects

Flooding can result in severe devastation of nearby areas. Flash floods and tsunamis can result in sweeping waters that travel at destructive speeds. Fast-moving water has the power to demolish all obstacles in its path such as homes, trees, bridges, and buildings. Animals, plants, and humans may all lose their lives during a flood.

Floods can also cause pollution and infection. Sewage may seep from drains or septic tanks and contaminate drinking water or surrounding lands. Similarly, toxins, fuels, debris from annihilated buildings, and other hazardous materials can leave water unusable for consumption. (38) As the water begins to drain, mold may begin to grow. As a result, residents of flooded areas may be left without power, drinkable water, or be exposed to toxins and other diseases.

(39) Although often associated with devastation, not all flooding results in adverse circumstances. For thousands of years, peoples have inhabited floodplains of rivers. (41) Examples include the Mississippi Valley of the United States, the Nile River in Egypt, and the Tigris River of the Middle East. The flooding of such rivers (42) caused nutrient-rich silts to be deposited on the floodplains. Thus, after the floods recede, an extremely fertile soil is left behind. This soil is conducive to the agriculture of bountiful crops and has sustained the diets of humans for millennium.

Proactive Measures Against Flooding

Technologies now allow scientists to predict where and when flooding is likely to occur. Such technologies can also be used (43) to project the severity of an anticipated flood. In this way, local inhabitants can be warned and take preventative measures such as boarding up their homes, gathering necessary provisions, and moving themselves and possessions to higher grounds.

The (44) picturesque views of coastal regions and rivers have long enticed people to build near such locations. Due to the costs associated with the repairs needed after the flooding of such residencies, many governments now require inhabitants of flood-prone areas to purchase flood insurance and build flood-resistant structures. Pictures of all items within a building or home should be taken so that proper reimbursement for losses can be made in the event that a flood does occur.

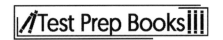

Staying Safe During a Flood

If a forecasted flood does occur, then people should retreat to higher ground such as a mountain or roof. Flooded waters may be contaminated, contain hidden debris, or travel at high speeds. Therefore, people should not attempt to walk or drive through a flooded area. To prevent electrocution, electrical outlets and downed power lines need to be avoided.

The Flood Dries Up

Regardless of the type or cause of a flood, floods can result in detrimental alterations to nearby lands and serious injuries to nearby inhabitants. By understanding flood cycles, civilizations can learn to take advantage of flood seasons. By taking the proper precautionary measures, people can stay safe when floods occur. Thus, proper knowledge can lead to safety and prosperity during such an adverse natural phenomenon.

34. What information from the graphs could be used to support the claims found in the third paragraph?
    a. Between 1970-1980, Boston experienced an increase in the number of days with flood events.
    b. Between 1970-1980, Atlantic City, New Jersey did not experience an increase in the number of days with flood events.
    c. Since 1970, the number of days with floods has decreased in major coastal cities across America.
    d. Since 1970, sea levels have risen along the East Coast.

35. One of the headings is entitled "A Floods After-Effects." How should this heading be rewritten?
    a. A Flood's After-Effect
    b. A Flood's After-Effects
    c. A Floods After-Affect
    d. A Flood's After-Affects

36. Which of the following revisions can be made to the sentence (reproduced below) that will still maintain the original meaning while making the sentence more concise?

    (36) Flooding can occur slowly or within seconds and can submerge small regions or extend over vast areas of land.

    a. NO CHANGE
    b. Flooding can either be slow or occur within seconds. It doesn't take long to submerge small regions or extend vast areas of land.
    c. Flooding occurs slowly or rapidly submerging vast areas of land.
    d. Vast areas of land can be flooded slowly or within seconds.

37. Which of the following would be the best choice for this sentence (reproduced below)?

They fear that the melting of icebergs will cause the (37) oceans levels to rise and flood coastal regions.

a. NO CHANGE
b. ocean levels
c. ocean's levels
d. levels of the oceans

38. Which choice best maintains the pattern of the first sentence of the paragraph?
a. NO CHANGE
b. As the rain subsides and the water begins to drain, mold may begin to grow.
c. Mold may begin to grow as the water begins to drain.
d. The water will begin to drain and mold will begin to grow.

39. Which of the following would be the best choice for this sentence (reproduced below)?

(39) Although often associated with devastation, not all flooding results in adverse circumstances. For thousands of years, peoples have inhabited floodplains of rivers.

a. NO CHANGE
b. Although often associated with devastation not all flooding results
c. Although often associated with devastation. Not all flooding results
d. While often associated with devastation, not all flooding results

40. What is the author's intent of the final paragraph?
a. To explain that all bad occurrences eventually come to an end.
b. To summarize the key points within the passage.
c. To explain that, with time, all flooded lands will eventually dry.
d. To relay a final key point about floods.

41. The author is considering deleting this sentence (reproduced below) from the tenth paragraph. Should the sentence be kept or deleted?

(41) Examples include the Mississippi Valley of the United States, the Nile River in Egypt, and the Tigris River of the Middle East.

a. Kept, because it provides examples of floodplains that have been successfully inhabited by civilizations.
b. Kept, because it provides an example of how floods can be beneficial.
c. Deleted, because it blurs the paragraph's focus on the benefits of floods.
d. Deleted, because it distracts from the overall meaning of the paragraph.

42. Which of the following would be the best choice for this sentence (reproduced below)?

The flooding of such rivers (42) caused nutrient-rich silts to be deposited on the floodplains.

   a. NO CHANGE
   b. Cause
   c. Causing
   d. Causes

43. Which of the following would be the best choice for this sentence (reproduced below)?

Such technologies can also be used (43) to project the severity of an anticipated flood.

   a. NO CHANGE
   b. Projecting
   c. Project
   d. Projected

44. Which term could best replace the underlined word (reproduced below)?

The (44) picturesque views of coastal regions and rivers have long enticed people to build near such locations.

   a. Colorful
   b. Drab
   c. Scenic
   d. Candid

Mathematics Test

1. If $6t + 4 = 16$, what is t?
   a. 1
   b. 2
   c. 3
   d. 4

2. The variable y is directly proportional to x. If $y = 3$ when $x = 5$, then what is y when $x = 20$?

3. A line passes through the point (1, 2) and crosses the y-axis at $y = 1$. Which of the following is an equation for this line?
   a. $y = 2x$
   b. $y = x + 1$
   c. $x + y = 1$
   d. $y = \frac{x}{2} - 2$

4. There are $4x + 1$ treats in each party favor bag. If a total of $60x + 15$ treats are distributed, how many bags are given out?
   a. 15
   b. 16
   c. 20
   d. 22

5. Apples cost $2 each, while oranges cost $3 each. Maria purchased 10 fruits in total and spent $22. How many apples did she buy?

6. What are the polynomial roots of $x^2 + x - 2$?
    a. 1 and 2
    b. -1 and 2
    c. 2 and -2
    d. 9 and 13

7. What is the y-intercept of $y = x^{5/3} + (x - 3)(x + 1)$?
    a. 3.5
    b. 7.6
    c. -3
    d. -15.1

8. $x^4 - 16$ can be simplified to which of the following?
    a. $(x^2 - 4)(x^2 + 4)$
    b. $(x^2 + 4)(x^2 + 4)$
    c. $(x^2 - 4)(x^2 - 4)$
    d. $(x^2 - 2)(x^2 + 4)$

9. $(4x^2y^4)^{\frac{3}{2}}$ can be simplified to which of the following?
    a. $8x^3y^6$
    b. $4x^{\frac{5}{2}}y$
    c. $4xy$
    d. $32x^{\frac{7}{2}}y^{\frac{11}{2}}$

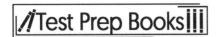

10. If $\sqrt{1+x} = 4$, what is x?

11. Suppose $\frac{x+2}{x} = 2$. What is x?

 a. -1
 b. 0
 c. 2
 d. 4

12. A ball is thrown from the top of a high hill, so that the height of the ball as a function of time is $h(t) = -16t^2 + 4t + 6$, in feet. What is the maximum height of the ball in feet?

 a. 6
 b. 6.25
 c. 6.5
 d. 6.75

13. A rectangle has a length that is 5 feet longer than three times its width. If the perimeter is 90 feet, what is the length in feet?

 a. 10
 b. 20
 c. 25
 d. 35

14. Five students take a test. The scores of the first four students are 80, 85, 75, and 60. If the median score is 80, which of the following could NOT be the score of the fifth student?

 a. 60
 b. 80
 c. 85
 d. 100

15.

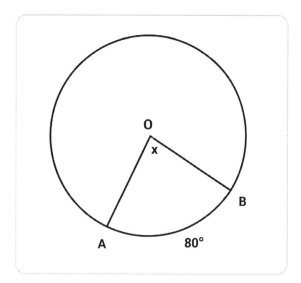

The area of circle O is 49π m. What is the area of the sector formed by ∠ *AOB*?

    a. 80π m

    b. 10.9π m

    c. 4.9π m

    d. 10π m

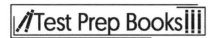
16. The triangle shown below is a right triangle. What's the value of x?

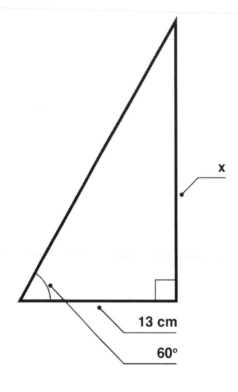

a. $x = 1.73$
b. $x = 0.57$
c. $x = 13$
d. $x = 22.49$

17. A company invests $50,000 in a building where they can produce saws. If the cost of producing one saw is $40, then which function expresses the amount of money the company pays? The variable y is the money paid and x is the number of saws produced.
a. $y = 50,000x + 40$
b. $y + 40 = x - 50,000$
c. $y = 40x - 50,000$
d. $y = 40x + 50,000$

18. A six-sided die is rolled. What is the probability that the roll is 1 or 2?
a. $\frac{1}{6}$

b. $\frac{1}{4}$

c. $\frac{1}{3}$

d. $\frac{1}{2}$

19. A line passes through the origin and through the point (-3, 4). What is the slope of the line?

    a. $-\dfrac{4}{3}$

    b. $-\dfrac{3}{4}$

    c. $\dfrac{4}{3}$

    d. $\dfrac{3}{4}$

20. A pair of dice is thrown, and the sum of the two scores is calculated. What's the expected value of the roll?

    a. 5
    b. 6
    c. 7
    d. 8

21.

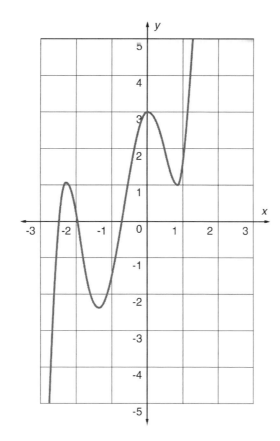

Which of the following functions represents the graph above?

    a. $y = x^5 + 3.5x^4 - 6.5x^2 + 0.5x + 3$
    b. $y = x^5 - 3.5x^4 + 6.5x^2 - 0.5x - 3$
    c. $y = 5x^4 - 6.5x^2 + 0.5x + 3$
    d. $y = -5x^4 - 6.5x^2 + 0.5x + 3$

22. Katie works at a clothing company and sold 192 shirts over the weekend. One third of the shirts that were sold were patterned, and the rest were solid. Which mathematical expression would calculate the number of solid shirts Katie sold over the weekend?

    a. $192 \times \frac{1}{3}$

    b. $192 \div \frac{1}{3}$

    c. $192 \times (1 - \frac{1}{3})$

    d. $192 \div 3$

23. For a group of 20 men, the median weight is 180 pounds and the range is 30 pounds. If each man gains 10 pounds, which of the following would be true?

    a. The median weight will increase, and the range will remain the same.
    b. The median weight and range will both remain the same.
    c. The median weight will stay the same, and the range will increase.
    d. The median weight and range will both increase.

24. Given the value of a given stock at monthly intervals, which graph should be used to best represent the trend of the stock?

    a. Box plot
    b. Line plot
    c. Line graph
    d. Circle graph

25. What is the probability of randomly picking the winner and runner-up from a race of 4 horses and distinguishing which is the winner?

    a. $\frac{1}{4}$

    b. $\frac{1}{2}$

    c. $\frac{1}{16}$

    d. $\frac{1}{12}$

27. A shipping box has a length of 8 inches, a width of 14 inches, and a height of 4 inches. If all three dimensions are doubled, what is the relationship between the volume of the new box and the volume of the original box?

    a. The volume of the new box is double the volume of the original box.
    b. The volume of the new box is four times as large as the volume of the original box.
    c. The volume of the new box is six times as large as the volume of the original box.
    d. The volume of the new box is eight times as large as the volume of the original box.

29.

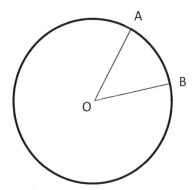

The length of arc $AB = 3\pi$ cm. The length of $\overline{OA} = 12$ cm. What is the degree measure of $\angle AOB$?

    a. 30 degrees
    b. 40 degrees
    c. 45 degrees
    d. 55 degrees

30. If x is not zero, then $\dfrac{3}{x} + \dfrac{5u}{2x} - \dfrac{u}{4} =$

    a. $\dfrac{12+10u-ux}{4x}$

    b. $\dfrac{3+5u-ux}{x}$

    c. $\dfrac{12x+10u+ux}{4x}$

    d. $\dfrac{12+10u-u}{4x}$

31. Which of the following inequalities is equivalent to $3 - \dfrac{1}{2}x \geq 2$?

    a. $x \geq 2$
    b. $x \leq 2$
    c. $x \geq 1$
    d. $x \leq 1$

32. If $g(x) = x^3 - 3x^2 - 2x + 6$ and $f(x) = 2$, then what is $g(f(x))$?

    a. -26
    b. 6
    c. $2x^3 - 6x^2 - 4x + 12$
    d. -2

33. What are the coordinates of the focus of the parabola $y = -9x^2$?

    a. $(-3, 0)$

    b. $\left(-\frac{1}{36}, 0\right)$

    c. $(0, -3)$

    d. $\left(0, -\frac{1}{36}\right)$

34. A National Hockey League store in the state of Michigan advertises 50% off all items. Sales tax in Michigan is 6%. How much would a hat originally priced at $32.99 and a jersey originally priced at $64.99 cost during this sale? Round to the nearest penny.

    a. $97.98

    b. $103.86

    c. $51.93

    d. $48.99

35. Store brand coffee beans cost $1.23 per pound. A local coffee bean roaster charges $1.98 per 1 ½ pounds. How much more would 5 pounds from the local roaster cost than 5 pounds of the store brand?

    a. $0.55

    b. $1.55

    c. $1.45

    d. $0.45

36. Which equation best represents the scatterplot below?

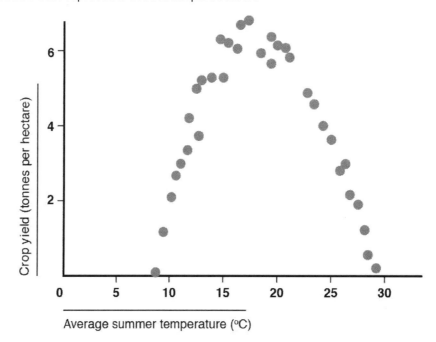

    a. $y = 3x - 4$

    b. $y = 2x^2 + 7x - 9$

    c. $y = (3)(4^x)$

d. $y = -\frac{1}{14}x^2 + 2x - 8$

37. What is the volume of a cylinder, in terms of π, with a radius of 5 inches and a height of 10 inches?
    a. 250 π in³
    b. 50 π in³
    c. 100 π in³
    d. 200 π in³

38. What is the solution to the following system of equations?
$$x^2 - 2x + y = 8$$
$$x - y = -2$$

    a. $(-2, 3)$
    b. There is no solution.
    c. $(-2, 0)$ $(1, 3)$
    d. $(-2, 0)$ $(3, 5)$

No Calculator Questions

39. Which of the ordered pairs below is a solution to the following system of inequalities?
$$y > 2x \quad 3$$
$$y < -4x + 8$$

    a. $(4, 5)$
    b. $(-3, -2)$
    c. $(3, -1)$
    d. $(5, 2)$

40. In an office, there are 50 workers. A total of 60% of the workers are women, and the chances of a woman wearing a skirt is 50%. If no men wear skirts, how many workers are wearing skirts?

41. Ten students take a test. Five students get a 50. Four students get a 70. If the average score is 55, what was the last student's score?

a. 20
b. 40
c. 50
d. 60

42. What's the probability of rolling a 6 at least once in two rolls of a die?

a. $\frac{1}{3}$

b. $\frac{1}{36}$

c. $\frac{1}{6}$

d. $\frac{11}{36}$

43. A solution needs 5 mL of saline for every 8 mL of medicine given. How much saline is needed for 45 mL of medicine?

a. $\frac{225}{8}$ mL

b. 72 mL

c. 28 mL

d. $\frac{45}{8}$ mL

44. What's the midpoint of a line segment with endpoints $(-1, 2)$ and $(3, -6)$?

a. $(1, 2)$
b. $(1, 0)$
c. $(-1, 2)$
d. $(1, -2)$

45. An equilateral triangle has a perimeter of 18 feet. If a square whose sides have the same length as one side of the triangle is built, what will be the area of the square?
    a. 6 square feet
    b. 36 square feet
    c. 256 square feet
    d. 1000 square feet

46. What is the volume of a sphere, in terms of π, with a radius of 3 inches?
    a. $36\,\pi$ in³
    b. $27\,\pi$ in³
    c. $9\,\pi$ in³
    d. $72\,\pi$ in³

47. What is the length of the other leg of a right triangle with a hypotenuse of 10 inches and a leg of 8 inches?

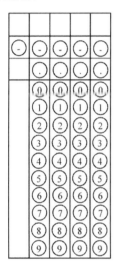

48. A pizzeria owner regularly creates jumbo pizzas, each with a radius of 9 inches. She is mathematically inclined, and wants to know the area of the pizza to purchase the correct boxes and know how much she is feeding her customers. What is the area of the circle, in terms of π, with a radius of 9 inches?
    a. $81\,\pi$ in²
    b. $18\,\pi$ in²
    c. $90\,\pi$ in²
    d. $9\,\pi$ in²

49. How will the following number be written in standard form:
$$(1 \times 10^4) + (3 \times 10^3) + (7 \times 10^1) + (8 \times 10^0)$$
    a. 137
    b. 13,078
    c. 1,378
    d. 8,731

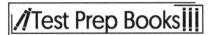

50. What is the simplified form of the expression $tan\theta \, cos\theta$?

    a. $sin\theta$

    b. 1

    c. $csc\theta$

    d. $\dfrac{1}{sec\theta}$

51. What is the value of the sum of $\dfrac{1}{3}$ and $\dfrac{2}{5}$?

    a. $\dfrac{3}{8}$

    b. $\dfrac{11}{15}$

    c. $\dfrac{11}{30}$

    d. $\dfrac{4}{5}$

52. If the cosine of $30° = x$, the sine of what angle also equals x?

    a. 30°

    b. 60°

    c. 90°

    d. 120°

53. What is the next number in the following series: $1, 3, 6, 10, 15, 21, \dots$?

54.

If $\overline{AE} = 4$, $\overline{AB} = 5$, and $\overline{AD} = 5$, what is the length of $\overline{AC}$?

⊖	⊖	⊖	⊖	⊖
	·	·	·	·
0	0	0	0	
1	1	1	1	
2	2	2	2	
3	3	3	3	
4	4	4	4	
5	5	5	5	
6	6	6	6	
7	7	7	7	
8	8	8	8	
9	9	9	9	

55. What is the next number in the following series: 1, 3, 6, 10, 15, 21, ... ?

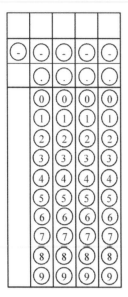

56. 6 is 30% of what number?

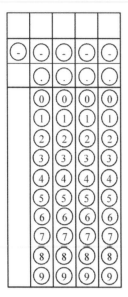

57. What is the value of the following expression?

$$\sqrt{8^2 + 6^2}$$

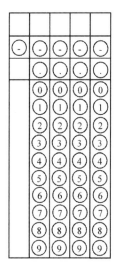

58. What is the measurement of angle f in the following picture? Assume the lines are parallel.

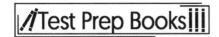

Essay Prompt

For-profit institutions of learning should be illegal because they suffer from a conflict of interest between the student receiving the best education and the institution minimizing costs in providing instruction and other services to remain profitable.

Write a response in support or nonsupport of the statement in regard to the position that is taken. To support your stance, evaluate how your position may or may not be true to the argument. Explain how you've come to your position and provide examples in support.

Answer Explanations #1

Reading Test

1. C: Gulliver becomes acquainted with the people and practices of his new surroundings. Choice *C* is the correct answer because it most extensively summarizes the entire passage. While Choices *A* and *B* are reasonable possibilities, they reference portions of Gulliver's experiences, not the whole. Choice *D* is incorrect because Gulliver doesn't express repentance or sorrow in this particular passage.

2. A: Principal refers to *chief* or *primary* within the context of this text. Choice *A* is the answer that most closely aligns with this definition. Choices *B* and *D* make reference to a helper or followers while Choice *C* doesn't meet the description of Reldresal from the passage.

3. C: One can reasonably infer that Gulliver is considerably larger than the children who were playing around him because multiple children could fit into his hand. Choice *B* is incorrect because there is no indication of stress in Gulliver's tone. Choices *A* and *D* aren't the best answer because though Gulliver seems fond of his new acquaintances, he didn't travel there with the intentions of meeting new people or to express a definite love for them in this particular portion of the text.

4. C: The emperor made a *definitive decision* to expose Gulliver to their native customs. In this instance, the word *mind* was not related to a vote, question, or cognitive ability.

5. A: Choice *A* is correct. This assertion does *not* support the fact that games are a commonplace event in this culture because it mentions conduct, not games. Choices *B*, *C*, and *D* are incorrect because these do support the fact that games were a commonplace event.

6. B: Choice *B* is the only option that mentions the correlation between physical ability and leadership positions. Choices *A* and *D* are unrelated to physical strength and leadership abilities. Choice *C* does not make a deduction that would lead to the correct answer—it only comments upon the abilities of common townspeople.

7. D: It emphasizes Mr. Utterson's anguish in failing to identify Hyde's whereabouts. Context clues indicate that Choice *D* is correct because the passage provides great detail of Mr. Utterson's feelings about locating Hyde. Choice *A* does not fit because there is no mention of Mr. Lanyon's mental state. Choice *B* is incorrect; although the text does make mention of bells, Choice *B* is not the *best* answer overall. Choice *C* is incorrect because the passage clearly states that Mr. Utterson was determined, not unsure.

8. A: In the city. The word *city* appears in the passage several times, thus establishing the location for the reader.

9. B: It scares children. The passage states that the Juggernaut causes the children to scream. Choices *A* and *D* don't apply because the text doesn't mention either of these instances specifically. Choice *C* is incorrect because there is nothing in the text that mentions space travel.

10. B: To constantly visit. The mention of *morning*, *noon*, and *night* make it clear that the word *haunt* refers to frequent appearances at various times. Choice *A* doesn't work because the text makes no mention of levitating. Choices *C* and *D* are not correct because the text makes mention of Mr. Utterson's anguish and disheartenment because of his failure to find Hyde but does not make mention of Mr. Utterson's feelings negatively affecting anyone else.

11. D: This is an example of alliteration. Choice *D* is the correct answer because of the repetition of the *L*-words. Hyperbole is an exaggeration, so Choice *A* doesn't work. No comparison is being made, so no simile or juxtaposition is being used, thus eliminating Choices B and C.

12. D: The speaker intends to continue to look for Hyde. Choices *A* and *B* are not possible answers because the text doesn't refer to any name changes or an identity crisis, despite Mr. Utterson's extreme obsession with finding Hyde. The text also makes no mention of a mistaken identity when referring to Hyde, so Choice *C* is also incorrect.

13. A: The tone is exasperated. While contemplative is an option because of the inquisitive nature of the text, Choice *A* is correct because the speaker is frustrated by the thought of being included when he felt that the fellow members of his race were being excluded. The speaker is not nonchalant, nor accepting of the circumstances which he describes.

14. C: Choice *C*, *contented*, is the only word that has different meaning. Furthermore, the speaker expresses objection and disdain throughout the entire text.

15. B: To address the hypocrisy of the Fourth of July holiday. While the speaker makes biblical references, it is not the main focus of the passage, thus eliminating Choice *A* as an answer. The passage also makes no mention of wealthy landowners and doesn't speak of any positive response to the historical events, so Choices *C* and *D* are not correct.

16. D: Choice *D* is the correct answer because it clearly makes reference to justice being denied.

17. D: Hyperbole. Choices *A* and *B* are unrelated. Assonance is the repetition of sounds and commonly occurs in poetry. Parallelism refers to two statements that correlate in some manner. Choice *C* is incorrect because amplification normally refers to clarification of meaning by broadening the sentence structure, while hyperbole refers to a phrase or statement that is being exaggerated.

18. D: "Rivers of Babylon" and "songs of Zion" are considered allusions. Specifically, they are biblical allusions. Allusions are a historical or literary reference to a person, place, or thing. Choice *A*, simile, is a comparative phrase that uses "like" or "as," so this is incorrect. Choice *B*, metaphor, is a comparison that says one thing *is* another thing, such as "her voice is the sun." Choice *C*, personification, is giving human characteristics to an inanimate object or animal, so this is also incorrect.

19. B: A period of time. It is apparent that Lincoln is referring to a period of time within the context of the passage because of how the sentence is structured with the word *ago*.

20. C: Lincoln's reference to *the brave men, living and dead, who struggled here,* proves that he is referring to a battlefield. Choices *A* and *B* are incorrect, as a *civil war* is mentioned and not a war with France or a war in the Sahara Desert. Choice *D* is incorrect because it does not make sense to consecrate a President's ground instead of a battlefield ground for soldiers who died during the American Civil War.

21. D: Abraham Lincoln is a former president of the United States, and he referenced a "civil war" during his address.

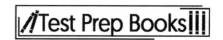

22. A: The audience should perpetuate the ideals of freedom that the soldiers died fighting for. Lincoln doesn't address any of the topics outlined in Choices *B*, *C*, or *D*. Therefore, Choice *A* is the correct answer.

23. D: Choice *D*, epiphora (also called epistrophe), is the correct answer because of the repetition of the word *people* at the end of the passage. Choice *A*, *antimetabole*, is the repetition of words in a phrase or clause but in reverse order, such as: "I do what I like, and like what I do." Choice *B*, *antiphrasis*, is a form of denial of an assertion in a text. Choice *C*, *anaphora*, is the repetition that occurs at the beginning of sentences.

24. A: Choice *A* is correct because Lincoln's intention was to memorialize the soldiers who had fallen as a result of war as well as celebrate those who had put their lives in danger for the sake of their country. Choices *B* and *D* are incorrect because Lincoln's speech was supposed to foster a sense of pride among the members of the audience while connecting them to the soldiers' experiences.

25. A: The word *sponsorship* is closely related to the word *patronage*, as both words refer to a type of aid received by someone who is lending support. Choice *B*, aberration, means deviating from the right or normal course, so this is incorrect. Choice *C*, caustic, is an adjective and means capable of burning or corroding, so this is incorrect. Choice *D*, adulation, means excessive flattery or devotion to someone, so this is also incorrect.

26. D: *Working man* is most closely aligned with Choice *D*, *bourgeois.* In the context of the speech, the word *bourgeois* means *working* or *middle class*. Choice *A*, *Plebeian*, does suggest *common people*; however, this is a term that is specific to ancient Rome. Choice *B*, *viscount*, is a European title used to describe a specific degree of nobility. Choice *C*, *entrepreneur*, is a person who operates their own business.

27. C: In the context of the speech, the term *working man* most closely correlates with Choice *C*, *working man is someone who works for wages among the middle class.* Choice *A* is not mentioned in the passage and is off-topic. Choice *B* may be true in some cases, but it does not reflect the sentiment described for the term *working man* in the passage. Choice *D* may also be arguably true. However, it is not given as a definition but as *acts* of the working man, and the topics of *field, factory,* and *screen* are not mentioned in the passage.

28. D: *Enterprise* most closely means *cause.* Choices *A, B,* and *C* are all related to the term *enterprise.* However, Dickens speaks of a *cause* here, not a company, courage, or a game. *He will stand by such an enterprise* is a call to stand by a cause to enable the working man to have a certain autonomy over his own economic standing. The very first paragraph ends with the statement that the working man *shall . . . have a share in the management of an institution which is designed for his benefit.*

29. B: The speaker's salutation is one from an entertainer to his audience and uses the friendly language to connect to his audience before a serious speech. Recall in the first paragraph that the speaker is there to "accompany [the audience] . . . through one of my little Christmas books," making him an author there to entertain the crowd with his own writing. The speech preceding the reading is the passage itself, and, as the tone indicates, a serious speech addressing the "working man." Although the passage speaks of employers and employees, the speaker himself is not an employer of the audience, so Choice *A* is incorrect. Choice *C* is also incorrect, as the salutation is not used ironically, but sincerely, as the speech addresses the wellbeing of the crowd. Choice *D* is incorrect because the speech is not given by a politician, but by a writer.

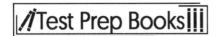

30. B: For the working man to have a say in his institution which is designed for his benefit. Choice *A* is incorrect because that is the speaker's *first* desire, not his second. Choices *C* and *D* are tricky because the language of both of these is mentioned after the word *second*. However, the speaker doesn't get to the second wish until the next sentence. Choices *C* and *D* are merely prepositions preparing for the statement of the main clause, Choice *B*.

31. D: The use of "I" could serve to have a "hedging" effect, allow the reader to connect with the author in a more personal way, and cause the reader to empathize more with the egrets. However, it doesn't distance the reader from the text, making Choice *D* the answer to this question.

32. C: The quote provides an example of a warden protecting one of the colonies. Choice *A* is incorrect because the speaker of the quote is a warden, not a hunter. Choice *B* is incorrect because the quote does not lighten the mood, but shows the danger of the situation between the wardens and the hunters. Choice *D* is incorrect because there is no humor found in the quote.

33. D: A *rookery* is a colony of breeding birds. Although *rookery* could mean Choice *A*, houses in a slum area, it does not make sense in this context. Choices *B* and *C* are both incorrect, as this is not a place for hunters to trade tools or for wardens to trade stories.

34. B: An important bird colony. The previous sentence is describing "twenty colonies" of birds, so what follows should be a bird colony. Choice *A* may be true, but we have no evidence of this in the text. Choice *C* does touch on the tension between the hunters and wardens, but there is no official "Bird Island Battle" mentioned in the text. Choice *D* does not exist in the text.

35. D: To demonstrate the success of the protective work of the Audubon Association. The text mentions several different times how and why the association has been successful and gives examples to back this fact. Choice *A* is incorrect because although the article, in some instances, calls certain people to act, it is not the purpose of the entire passage. There is no way to tell if Choices *B* and *C* are correct, as they are not mentioned in the text.

36. C: To have a better opportunity to hunt the birds. Choice *A* might be true in a general sense, but it is not relevant to the context of the text. Choice *B* is incorrect because the hunters are not studying lines of flight to help wardens, but to hunt birds. Choice *D* is incorrect because nothing in the text mentions that hunters are trying to build homes underneath lines of flight of birds for good luck.

37. A: It introduces certain insects that transition from water to air. Choice *B* is incorrect because although the passage talks about gills, it is not the central idea of the passage. Choices *C* and *D* are incorrect because the passage does not "define" or "invite," but only serves as an introduction to stoneflies, dragonflies, and mayflies and their transition from water to air.

38. C: The act of shedding part or all of the outer shell. Choices *A*, *B*, and *D* are incorrect.

39. B: The first paragraph serves as a contrast to the second. Notice how the first paragraph goes into detail describing how insects are able to breathe air. The second paragraph acts as a contrast to the first by stating "[i]t is of great interest to find that, nevertheless, a number of insects spend much of their time under water." Watch for transition words such as "nevertheless" to help find what type of passage you're dealing with.

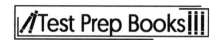

40: C: The stage of preparation in between molting is acted out in the water, while the last stage is in the air. Choices *A, B,* and *D* are all incorrect. *Instars* is the phase between two periods of molting, and the text explains when these transitions occur.

41. C: The author's tone is informative and exhibits interest in the subject of the study. Overall, the author presents us with information on the subject. One moment where personal interest is depicted is when the author states, "It is of great interest to find that, nevertheless, a number of insects spend much of their time under water."

42. C: Their larva can breathe the air dissolved in water through gills of some kind. This is stated in the last paragraph. Choice *A* is incorrect because the text mentions this in a general way at the beginning of the passage concerning "insects as a whole." Choice *B* is incorrect because this is stated of beetles and water-bugs, and not the insects in question. Choice *D* is incorrect because this is the opposite of what the text says of instars.

43. D: To enlighten the audience on the habits of sun-fish and their hatcheries. Choice *A* is incorrect because although the Adirondack region is mentioned in the text, there is no cause or effect relationships between the region and fish hatcheries depicted here. Choice *B* is incorrect because the text does not have an agenda, but rather is meant to inform the audience. Finally, Choice *C* is incorrect because the text says nothing of how sun-fish mate.

44. B: The word *wise* in this passage most closely means *manner*. Choices *A* and *C* are synonyms of *wise*; however, they are not relevant in the context of the text. Choice *D*, *ignorance*, is opposite of the word *wise* and is therefore incorrect.

45. A: Fish at the stage of development where they are capable of feeding themselves. Even if the word *fry* isn't immediately known to the reader, the context gives a hint when it says "until the fry are hatched out and are sufficiently large to take charge of themselves."

46. B: The sun-fish builds it with her tail and snout. The text explains this in the second paragraph: "she builds, with her tail and snout, a circular embankment 3 inches in height and 2 thick." Choice *A* is used in the text as a simile.

47. D: To conclude a sequence and add a final detail. The concluding sequence is expressed in the phrase "[t]he mother sun-fish, having now built or provided her 'hatchery.'" The final detail is the way in which the sun-fish guards the "inclosure." Choices *A, B,* and *C* are incorrect.

48. C: *Extraneous* most nearly means *superfluous*, or *trivial*. Choice *A*, *indispensable*, is incorrect because it means the opposite of *extraneous*. Choice *B*, *bewildering*, means *confusing* and is not relevant to the context of the sentence. Finally, Choice *D* is wrong because although the prefix of the word is the same, *ex-*, the word *exuberant* means *elated* or *enthusiastic*, and is irrelevant to the context of the sentence.

49. A: Bring to light an alternative view on human perception by examining the role of technology in human understanding. This is a challenging question because the author's purpose is somewhat open-ended. The author concludes by stating that the questions regarding human perception and observation can be approached from many angles. Thus they do not seem to be attempting to prove one thing or another. Choice *B* is incorrect because we cannot know for certain whether the electron experiment is the latest discovery in astroparticle physics because no date is given. Choice *C* is a broad generalization that does not reflect accurately on the writer's views. While the author does appear to reflect on opposing views of human understanding (Choice *D*), the best answer is Choice *A*.

50. C: It presents a problem, explains the details of that problem, and then ends with more inquiry. The beginning of this paragraph literally "presents a conundrum," explains the problem of partial understanding, and then ends with more questions, or inquiry. There is no solution offered in this paragraph, making Choices A and B incorrect. Choice D is incorrect because the paragraph does not begin with a definition.

51. D: Looking back in the text, the author describes that classical philosophy holds that understanding can be reached by careful observation. This will not work if they are overly invested or biased in their pursuit. Choices A and C are in no way related and are completely unnecessary. A specific theory is not necessary to understanding, according to classical philosophy mentioned by the author. Again, the key to understanding is observing the phenomena outside of it, without bias or predisposition. Thus, Choice B is wrong.

52. B: The electrons passed through both holes and then onto the plate. Choices A and C are wrong because such movement is not mentioned at all in the text. In the passage, the author says that electrons that were physically observed appeared to pass through one hole or another. Remember, the electrons that were observed doing this were described as acting like particles. Therefore, Choice D is wrong. Recall that the plate actually recorded electrons passing through both holes simultaneously and hitting the plate. This behavior, the electron activity that wasn't seen by humans, was characteristic of waves. Thus, Choice B is the right answer.

Writing and Language Test

1. D: This passage is informative (D) because it is nonfiction and factual. The passage's intent is not to state an opinion, discuss an individual's life, or tell a story. Thus, the passage is not argumentative (A), biographical (C), or narrative (B).

2. D: To begin, *of* is not required here. *Apprenticeship* is also more appropriate in this context than *apprentice opportunities*, *apprentice* describes an individual in an apprenticeship, not an apprenticeship itself. Both of these changes are needed, making (D) the correct answer.

3. D: To begin, the selected sentence is a run-on, and displays confusing information. Thus, the sentence does need revision, making (A) wrong. The main objective of the selected section of the passage is to communicate that many positions (*positions* is a more suitable term than *offices,* as well) require a PE license, which is gained by scoring well on the FE and PE assessments. This must be the primary focus of the revision. It is necessary to break the sentence into two, to avoid a run-on. Choice B fixes the run-on aspect, but the sentence is indirect and awkward in construction. It takes too long to establish the importance of the PE license. Choice C is wrong for the same reason and it is a run on. Choice D is correct because it breaks the section into coherent sentences and emphasizes the main point the author is trying to communicate: the PE license is required for some higher positions, it's obtained by scoring well on the two standardized assessments, and college and experience can be used to prepare for the assessments in order to gain the certification.

4. C: Any time a writer wants to validate a claim, he or she ought to provide factual information that proves or supports that claim: "beginning his or her preparation for the profession in high school" supports the claim that aircraft engineers undergo years of education. For this reason, Choice C is the correct response. However, completing such courses in high school does not guarantee that aircraft engineers will earn generous salaries (*A*), become employed in executive positions (*B*), or stay employed (*D*).

5. B: Choice *B* is correct because "skill" is defined as having certain aptitude for a given task. (*C*) is incorrect because "work" does not directly denote "critical thinking, business skills, problem solving, and creativity." (*A*) is incorrect because the word "care" doesn't fit into the context of the passage, and (*D*), "composition," is incorrect because nothing in this statement points to the way in which something is structured.

6. C: *Allows* is inappropriate because it does not stress what those in the position of aircraft engineers actually need to be able to do. *Requires* is the only alternative that fits because it actually describes necessary skills of the job.

7. B: The third paragraph discusses reports made by the United States Bureau of Labor Statistics (BLS) in regards to the median, upper 10 percent, and lower 10 percent annual salaries of aircraft engineers in 2015. Therefore, this paragraph is used to support the claim that aircraft engineers are compensated with generous salaries (*B*). The paragraph has nothing to do with an aircraft engineer's skill set (*A*), education (*C*), or incentive program (*D*).

8. A: The correct response is (*A*) because this statement's intent is to give examples as to how aircraft engineers apply mathematical equations and scientific processes towards aeronautical and aerospace issues and/or inventions. The answer is not "therefore" (*B*) or "furthermore" (*D*) because no causality is being made between ideas. Two items are neither being compared nor contrasted, so "however" (*C*) is also not the correct answer.

9. A: No change is required. The comma is properly placed after the introductory phrase "In May of 2015." Choice *B* is missing the word "in." Choice *C* does not separate the introductory phrase from the rest of the sentence. Choice *D* places an extra, and unnecessary, comma prior to 2015.

10. A: The word "conversely" best demonstrates the opposite sentiments in this passage. Choice *B* is incorrect because it denotes agreement with the previous statement. Choice *C* is incorrect because the sentiment is not restated but opposed. Choice *D* is incorrect because the previous statement is not a cause for the sentence in question.

11. A: Choice *A* is the correct answer because the projections are taking place in the present, even though they are making reference to a future date.

12. B: The passage contains clearly labeled subheadings. These subheadings inform the reader what will be addressed in upcoming paragraphs. Choice *A* is incorrect because the anti-terrorism laws of other countries were never addressed in the passage. The text is written in an informative manner; overly descriptive language is not utilized. Therefore, Choice *C* is incorrect. Choice *D* is incorrect because as mentioned, the structure of the text does help in the manner described in Choice *B*.

13. A: No change is needed. Choices *B* and *C* utilize incorrect comma placements. Choice *D* utilizes an incorrect verb tense (responding).

14. D: The third paragraph states "The majority of Muslims practice Islam peacefully". Therefore, the author explicitly states that most Muslims are peaceful peoples (*D*). Choices *B*, *C*, and *A* are not included in the passage and are incorrect.

15. B: The term "violate" implies a lack of respect or compliance. "Defile" means to degrade or show no respect. Therefore, (B) is the correct answer. Choice A is incorrect because "respect" is the opposite of violate. To "deny" is to refuse, so (C) is not the answer because the weight of the word "deny" is not as heavy as the word "violate." To "obey" is to follow orders, so (D) is not the answer.

16. B: An allusion is a direct or indirect literary reference or figure of speech towards a person, place, or event. By referencing the diversion of the airplanes to alternate locations, the author uses an allusion (Choice B) to highlight the impact of United Flight 93. Although a graph depicting the decline in the number of aircraft passengers is provided, an image is not. Therefore, Choice A is not the answer. The passage does not tell the story from a single passenger's point of view. Thus, Choice C and Choice D are not the answers.

17. C: All of the choices except (C) go with the flow of the original underlined portion of the sentence and communicate the same idea. Choice C, however, does not take into account the rest of the sentence and therefore, becomes awkward and incorrect.

18. B: Although "diverging" means to separate from the main route and go in a different direction, it is used awkwardly and unconventionally in this sentence. Therefore, Choice A is not the answer. Choice B is the correct answer because it implies that the passengers distracted the terrorists, which caused a change in the plane's direction. "Converging" (C) is incorrect because it implies that the plane met another in a central location. Although the passengers may have distracted the terrorists, they did not distract the plane. Therefore, Choice D is incorrect.

19. D: The graph shows the number of people (in millions) boarding United States' flights between 1996-2005. The first few months following the attacks, the passengers boarding U.S. flights dropped to around 50 million when before the attacks there were around 70 million passengers boarding flights. Therefore, the correct answer is Choice D. The graph does not show where the flights were redirected (B), the number of passengers that other countries received as a result of the redirected air travel (C), or the resulting flight schedule implications (Choice A).

20. B: The last paragraph explains that museums and monuments have been erected to honor those who died as a result of the attacks and those who risked their lives to save the injured. Thus, the paragraph serves to explain the lasting impact on America and honor those impacted by the event (B). The design of the museums and monuments are not described, so Choice A is incorrect. Choice C is incorrect because America's War on Terror was not discussed in the last paragraph. Choice D is incorrect, because although the previous location of the towers was converted into a park, this was not mentioned in the passage.

21. C: *Intended* means planned or meant to. *Intended* is a far better choice than *desired*, because it would communicate goals and strategy more than simply saying that Bush desired to do something. *Desired* communicates wishing or direct motive. Choices B and D have irrelevant meanings and wouldn't serve the sentence at all.

22. A: While (B) isn't necessarily wrong, it lacks the direct nature that the original sentence has. Also, by breaking up the sentences like this, the reader becomes confused because the connection between the Taliban's defeat and ongoing war is now separated by a second sentence that is not necessary. Choice C corrects this problem but the fluidity of the sentence is marred because of the awkward construction of the first sentence. Choice D begins well, but lacks the use of *was* before overthrown, which discombobulates the sentence. While *yet* provides an adequate transition for the next sentence, the term *however* is more appropriate. Thus, the original structure of the two sentences is correct, making Choice *A*, NO CHANGE, the correct answer.

23. A: The comma after *result* is necessary for the sentence structure, making it an imperative component. The original sentence is correct, making Choice *A* correct. For the reason just listed, Choice *B* is incorrect because it lacks the crucial comma that introduces a new idea. Choice *C* is incorrect because a colon is unnecessary, and Choice *D* is wrong because the addition of "of" is both unnecessary and incorrect when applied to the rest of the sentence.

24. C: To be "gifted" is to be talented. "Academically" refers to education. Therefore, Fred Hampton was intellectually talented, or intelligent (*C*). Choice *B* is incorrect because it refers to a level of energy or activity. Choice *A* is incorrect because "vacuous" means the opposite of being gifted academically. Choice *D* is incorrect because it refers to one's physical build and/or abilities.

25. C: The goal for this question is to select a sentence that not only affirms, or backs up, the selected statement, but could also appear after it and flow with the rest of the piece. Choice *A* is irrelevant to the sentence; just because new members earned scholarships this doesn't necessarily mean that this was testament of Hampton's leadership or that this actually benefitted the NAACP. Choice *B* is very compelling. If Hampton got an award for the increase in numbers, this could bolster the idea that he was the direct cause of the rise in numbers and that he was of great value to the organization. However, it does not say directly that he was the cause of the increase and that this was extremely beneficial to the NAACP. Choice *C* is a much better choice than Choice *B*. Choice *C* mentions that the increase in members is unprecedented. Because there has never been this large an increase before, it can be concluded that this increase was most likely due to Hampton's contributions. Thus, Choice *C* is correct. Choice *D* does nothing for the underlined section.

26. B: Choice *B* moves the word "eventually" to the beginning of the sentences. By using the term as an introductory word, continuity from one sentence to another is created. Meanwhile, the syntax is not lost. Choice *A* is incorrect because the sentence requires a proper transition. Choice *C* is incorrect because the sentence does not contain surprising or contrasting information, as is indicated by the introductory word "nevertheless." Choice *D* is incorrect -because the term "then" implies that Hampton's relocation to the BPP's headquarters in Chicago occurred shortly or immediately after leading the NAACP.

27. D: An individual with a charismatic personality is charming and appealing to others. Therefore, Choice *D* is the correct answer. Choice A is incorrect because someone with an egotistical personality is conceited or self-serving. Choice *B* is incorrect because "obnoxious" is the opposite of charismatic. Choice *C* is incorrect because someone with a chauvinistic personality is aggressive or prejudiced towards one's purpose, desire, or sex.

28. A: No change is needed: Choice *A*. The list of events accomplished by Hampton is short enough that each item in the list can be separated by a comma. Choice *B* is incorrect. Although a colon can be used to introduce a list of items, it is not a conventional choice for separating items within a series. Semicolons are used to separate at least three items in a series that have an internal comma. Semicolons can also be used to separate clauses in a sentence that contain internal commas intended for clarification purposes. Neither of the two latter uses of semicolons is required in the example sentence. Therefore, Choice *C* is incorrect. Choice *D* is incorrect because a dash is not a conventional choice for punctuating items in a series.

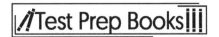

29. D: Claims can be supported with evidence or supporting details found within the text. Choice *D* is correct because Choices *A*, *B*, and *C* are all either directly stated or alluded to within the passage.

30. A: The term *neutralize* means to counteract, or render ineffective, which is exactly what the FBI is wanting to do. Accommodate means to be helpful or lend aid, which is the opposite of *neutralize*. Therefore (*B*) is wrong. *Assuage* means to ease, while *praise* means to express warm feeling, so they are in no way close to the needed context. Therefore, *neutralize* is the best option, making Choice *A*, NO CHANGE, the correct answer.

31. A: The original sentence suggests that the floor plans were provided to the FBI by O'Neal which facilitated the identification of the exact location of Hampton's bed. Choice *B* is incorrect because this states that the FBI provided O'Neal with the floor plans instead of the other way around. Choice *C* is incorrect because the sentence's word order conveys the meaning that O'Neal provided the FBI with Hampton's bed as well as the floor plans. Choice *D* is incorrect because it implies that it was the location of the bed that provided the FBI with the headquarters' floor plans.

32. B: "Commemorates" means to honor, celebrate, or memorialize a person or event. Therefore, Choice *B* is correct. Choice *A* is incorrect because "disregards" is the opposite of "commemorates." Choice *C* is incorrect because to communicate means to converse or to speak. Choice *D* is incorrect because to "deny" means to reject, negate, refuse, or rebuff.

33. D: From the context of the passage, it is clear that the author does not think well of the FBI and their investigation of Hampton and the Black Panthers. Choices *B* and *C* can be easily eliminated. "Well intended" is positive, which is not a characteristic that he would probably attribute to the FBI in the passage. Nor would he think they were "confused" but deliberate in their methods.. Choice *A*, "corrupt", is very compelling; he'd likely agree with this, but Choice *D*, "prejudiced" is better. The FBI may not have been corrupt but there certainly seemed to have particular dislike/distrust for the Black Panthers. Thus, Choice *D*, "prejudiced", is correct.

34. D: All of the cities included in the graphs are along the East Coast of the United States. All of the bars on the graphs show an increase in sea level or the number of days with flood events since 1970. Therefore, the author chose to include the graphs to support the claim that sea levels have risen along the East Coast since 1970, Choice *D*.

Choice *A* is incorrect because the bar above 1970 on Boston's graph is longer than the graphs' bar above 1980. Therefore, a. Between 1970-1980, Boston experienced an increase in the number of days with flood events. It's important to note that while there was a decrease from one decade to another, it does not negate the overall trend of increase in flooding events.

Choice *B* is incorrect because the bar above 1970 on Atlantic City's graph is shorter than the graphs' bar above 1980. Therefore, between 1970-1980, Atlantic City experienced an increase in the number of days with flood events. Choice *C* is incorrect because the bars increase in height on all of the cities' graphs, showing an increase in the number of days with floods along the entire East coast.

35. B: Although "affect" and "effect" sound the same, they have different meanings. "Affect" is used as a verb. It is defined as the influence of a person, place, or event on another. "Effect" is used as a noun. It is defined as the result of an event. Therefore, the latter ought to be used in the heading. For this reason, Choices *C* and *D* are incorrect. Because the effect is a result of the flood, a possessive

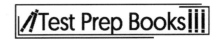

apostrophe is needed for the singular noun "flood." For this reason, Choice *A* is incorrect and Choice *B* is correct.

36. D: Again, the objective for questions like this is to determine if a revision is possible within the choices and if it can adhere to the specific criteria of the question; in this case, we want the sentence to maintain the original meaning while being more concise, or shorter. Choice *B* can be eliminated. The meaning of the original sentence is split into two distinct sentences. The second of the two sentences is also incorrectly constructed. Choice *C* is very intriguing but there is a jumble of verbs present in: "Flooding occurs slowly or rapidly submerging" that it makes the sentence awkward and difficult to understand without the use of a comma after *rapidly*, making it a poor construction. Choice *C* is wrong. Choice *D* is certainly more concise and it is correctly phrased; it communicates the meaning message that flooding can overtake great lengths of land either slowly or very fast. The use of "Vast areas of land" infers that smaller regions or small areas can flood just as well. Thus, Choice *D* is a good revision that maintains the meaning of the original sentence while being concise and more direct. This rules out Choice *A* in the process.

37. B: In this sentence, the word *ocean* does not require an *s* after it to make it plural because "ocean levels" is plural. Therefore (*A*) and (*C*) are incorrect. Because the passage is referring to multiple – if not all ocean levels – *ocean* does not require an apostrophe ('s) because that would indicate that only one ocean is the focus, which is not the case. Choice *D* does not fit well into the sentence and, once again, we see that *ocean* has an *s* after it. This leaves Choice *B*, which correctly completes the sentence and maintains the intended meaning.

38. C: Choice *C* is the best answer because it most closely maintains the sentence pattern of the first sentence of the paragraph, which begins with a noun and passive verb phrase. Choices *A* and *B* are incorrect because they do not maintain the sentence pattern of the first sentence of the paragraph. Instead, both choices add modifying prepositional phrases to the beginning of the sentence. Choice *D* is incorrect because it does not maintain the sentence pattern established by the first sentence of the paragraph. Instead, Choice *D* is an attempt to combine two independent clauses.

39. A: Choice *C* can be eliminated because creating a new sentence with *not* is grammatically incorrect and it throws off the rest of the sentence. Choice *B* is wrong because a comma is definitely needed after *devastation* in the sentence. Choice *D* is also incorrect because "while" is a poor substitute for "although". *Although* in this context is meant to show contradiction with the idea that floods are associated with devastation. Therefore, none of these choices would be suitable revisions because the original was correct: NO CHANGE, Choice *A*, is the correct answer.

40. B: Choice *B* is the correct answer because the final paragraph summarizes key points from each subsection of the text. Therefore, the final paragraph serves as the conclusion. A concluding paragraph is often found at the end of a text. It serves to remind the reader of the main points of a text. Choice *A* is incorrect because the last paragraph does not just mention adverse effects of floods. For example, the paragraph states "By understanding flood cycles, civilizations can learn to take advantage of flood seasons." Choice *C* is incorrect; although the subheading mentions the drying of floods, the phenomena is not mentioned in the paragraph. Finally, Choice *D* is incorrect because no new information is presented in the last paragraph of the passage.

41. A: Idea and claims are best expressed and supported within a text through examples, evidence, and descriptions. Choice *A* is correct because it provides examples of rivers that support the tenth paragraph's claim that "not all flooding results in adverse circumstances." Choice *B* is incorrect because the sentence does not explain how floods are beneficial. Therefore, Choices *C* and *D* are incorrect.

42. D: In the sentence, *caused* is an incorrect tense, making (*A*) wrong. Choice *B* is incorrect because this used as a noun, we need *cause* in verb form. Choices *C* and *D* are very compelling. *Causing* (*C*) is a verb and it is in the present continuous tense, which appears to agree with the verb flooding, but it is incorrectly used. This leaves (*D*), *causes*, which does fit because it is in the indefinite present tense. Fitting each choice into the sentence and reading it in your mind will also reveal that (*D*), *causes*, correctly completes the sentence. Apply this method to all the questions when possible.

43. A: To *project* means to anticipate or forecast. This goes very well with the sentence because it describes how new technology is trying to estimate flood activity in order to prevent damage and save lives. "Project" in this case needs to be assisted by "to" in order to function in the sentence. Therefore, Choice *A* is correct. Choices *B* and *D* are the incorrect tenses. Choice *C* is also wrong because it lacks *to*.

44. C: *Picturesque* is an adjective used for an attractive, scenic, or otherwise striking image. Thus, Choice *C* is correct. Choice *A* is incorrect because although "colorful" can be included in a picturesque view, it does not encompass the full meaning of the word. Choice *B* is incorrect because "drab" is the opposite of "picturesque." Choice *D* is incorrect because "candid" is defined as being frank, open, truthful, or honest.

Mathematics Test

1. B: First, subtract 4 from each side. This yields $6t = 12$. Now, divide both sides by 6 to obtain $t = 2$.

2.

To be directly proportional means that $y = mx$. If x is changed from 5 to 20, the value of x is multiplied by 4. Applying the same rule to the y-value, also multiply the value of y by 4. Therefore, $y = 12$.

3. B: From the slope-intercept form, $y = mx + b$, it is known that b is the y-intercept, which is 1. Compute the slope as $\frac{2-1}{1-0} = 1$, so the equation should be $y = x + 1$.

4. A: Each bag contributes $4x + 1$ treats. The total treats will be in the form $4nx + n$ where n is the total number of bags. The total is in the form $60x + 15$, from which it is known $n = 15$.

5.

Let a be the number of apples and o the number of oranges. Then, the total cost is $2a + 3o = 22$, while it also known that $a + o = 10$. Using the knowledge of systems of equations, cancel the o variables by multiplying the second equation by -3. This makes the equation:

$$-3a - 3o = -30$$

Adding this to the first equation, the b values cancel to get $-a = -8$, which simplifies to $a = 8$.

6. A: Finding the roots means finding the values of x when y is zero. The quadratic formula could be used, but in this case it is possible to factor by hand, since the numbers -1 and 2 add to 1 and multiply to -2. So, factor $x^2 + x - 2 = (x - 1)(x + 2) = 0$, then set each factor equal to zero. Solving for each value gives the values $x = 1$ and $x = -2$.

7. C: To find the y-intercept, substitute zero for x, which gives us:

$$y = 0^{\frac{5}{3}} + (0 - 3)(0 + 1)$$

$$0 + (-3)(1) = -3$$

8. A: This has the form $t^2 - y^2$, with $t = x^2$ and $y = 4$. It's also known that $t^2 - y^2 = (t + y)(t - y)$, and substituting the values for t and y into the right-hand side gives $(x^2 - 4)(x^2 + 4)$.

9. A: Simplify this to:

$$(4x^2y^4)^{\frac{3}{2}} = 4^{\frac{3}{2}}(x^2)^{\frac{3}{2}}(y^4)^{\frac{3}{2}}$$

Now:

$$4^{\frac{3}{2}} = (\sqrt{4})^3 = 2^3 = 8$$

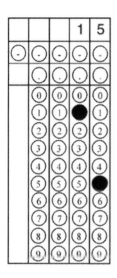
For the other, recall that the exponents must be multiplied, so this yields:

$$8x^{2 \cdot \frac{3}{2}} y^{4 \cdot \frac{3}{2}} = 8x^3 y^6$$

10.

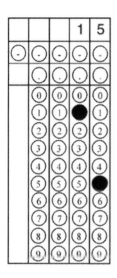

Start by squaring both sides to get $1 + x = 16$. Then subtract 1 from both sides to get $x = 15$.

11. C: Multiply both sides by x to get $x + 2 = 2x$, which simplifies to $x = 2$.

12. B: The independent variable's coordinate at the vertex of a parabola (which is the highest point, when the coefficient of the squared independent variable is negative) is given by $x = -\frac{b}{2a}$. Substitute and solve for x to get:

$$x = -\frac{4}{2(-16)} = \frac{1}{8}$$

Using this value of x, the maximum height of the ball (y), can be calculated. Substituting x into the equation yields:

$$h(t) = -16\frac{1^2}{8} + 4\frac{1}{8} + 6 = 6.25$$

13. D: Denote the width as w and the length as l. Then, $l = 3w + 5$. The perimeter is $2w + 2l = 90$. Substituting the first expression for l into the second equation yields:

$$2(3w + 5) + 2w = 90$$

$$6w + 10 + 2w = 90$$

$$8w = 80$$

$$w = 10$$

Putting this into the first equation, it yields:

$$l = 3(10) + 5 = 35$$

14. A: Lining up the given scores provides the following list: 60, 75, 80, 85, and one unknown. Because the median needs to be 80, it means 80 must be the middle data point out of these five. Therefore, the unknown data point must be the fourth or fifth data point, meaning it must be greater than or equal to 80. The only answer that fails to meet this condition is 60.

15. B: Given the area of the circle, the radius can be found using the formula $A = \pi r^2$. In this case, $49\pi = \pi r^2$, which yields $r = 7$ m. A central angle is equal to the degree measure of the arc it inscribes; therefore, $\angle x = 80°$. The area of a sector can be found using the formula:

$$A = \frac{\theta}{360°} \times \pi r^2$$

In this case:

$$A = \frac{80°}{360°} \times \pi(7)^2 = 10.9\pi \text{ m}$$

16. D: SOHCAHTOA is used to find the missing side length. Because the angle and adjacent side are known, $\tan 60 = \frac{x}{13}$. Making sure to evaluate tangent with an argument in degrees, this equation gives:

$$x = 13 \tan 60 = 13 \cdot 1.73 = 22.49$$

17. D: For manufacturing costs, there is a linear relationship between the cost to the company and the number produced, with a y-intercept given by the base cost of acquiring the means of production, and a slope given by the cost to produce one unit. In this case, that base cost is $50,000, while the cost per unit is $40. So, $y = 40x + 50,000$.

18. C: A die has an equal chance for each outcome. Since it has six sides, each outcome has a probability of $\frac{1}{6}$. The chance of a 1 or a 2 is therefore $\frac{1}{6} + \frac{1}{6} = \frac{1}{3}$.

19. A: The slope is given by:

$$m = \frac{y_2 - y_1}{x_2 - x_1} = \frac{0 - 4}{0 - (-3)} = -\frac{4}{3}$$

20. C: The expected value is equal to the total sum of each product of individual score and probability. There are 36 possible rolls. The probability of rolling a 2 is $\frac{1}{36}$. The probability of rolling a 3 is $\frac{2}{36}$. The probability of rolling a 4 is $\frac{3}{36}$. The probability of rolling a 5 is $\frac{4}{36}$. The probability of rolling a 6 is $\frac{5}{36}$. The probability of rolling a 7 is $\frac{6}{36}$. The probability of rolling an 8 is $\frac{5}{36}$. The probability of rolling a 9 is $\frac{4}{36}$. The probability of rolling a 10 is $\frac{3}{36}$. The probability of rolling an 11 is $\frac{2}{36}$. Finally, the probability of rolling a 12 is $\frac{1}{36}$.

Each possible outcome is multiplied by the probability of it occurring. Like this:

$$2 \times \frac{1}{36} = a$$

$$3 \times \frac{2}{36} = b$$

$$4 \times \frac{3}{36} = c$$

And so forth.

Then all of those results are added together:

$$a + b + c \ldots = expected\ value$$

In this case, it equals 7.

21. A: The graph contains four turning points (where the curve changes from rising to falling or vice versa). This indicates that the degree of the function (highest exponent for the variable) is 5, eliminating Choices C and D. The y-intercepts of the functions can be determined by substituting 0 for x and finding the value of y. The function for Choice A has a y-intercept of 3, and the function for Choice B has a y-intercept of -3. Therefore, Choice B is eliminated.

22. C: $\frac{1}{3}$ of the shirts sold were patterned. Therefore, $1 - \frac{1}{3} = \frac{2}{3}$ of the shirts sold were solid. Anytime "of" a quantity appears in a word problem, multiplication should be used. Therefore:

$$192 \times \frac{2}{3} = \frac{192 \times 2}{3} = \frac{384}{3} = 128 \text{ solid shirts were sold}$$

The entire expression is:

$$192 \times \left(1 - \frac{1}{3}\right)$$

23. A: If each man gains 10 pounds, every original data point will increase by 10 pounds. Therefore, the man with the original median will still have the median value, but that value will increase by 10. The smallest value and largest value will also increase by 10 and, therefore, the difference between the two won't change. The range does not change in value and, thus, remains the same.

24. C: Line graph. The scenario involves data consisting of two variables, month and stock value. Box plots display data consisting of values for one variable. Therefore, a box plot is not an appropriate choice. Both line plots and circle graphs are used to display frequencies within categorical data. Neither can be used for the given scenario. Line graphs display two numerical variables on a coordinate grid and show trends among the variables.

25. D: $\frac{1}{12}$. The probability of picking the winner of the race is $\frac{1}{4}$, or $\left(\frac{number\ of\ favorable\ outcomes}{number\ of\ total\ outcomes}\right)$. Assuming the winner was picked on the first selection, three horses remain from which to choose the

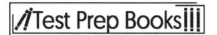

runner-up (these are dependent events). Therefore, the probability of picking the runner-up is $\frac{1}{3}$. To determine the probability of multiple events, the probability of each event is multiplied:

$$\frac{1}{4} \times \frac{1}{3} = \frac{1}{12}$$

27. D: The formula for finding the volume of a rectangular prism is $V = l \times w \times h$ where l is the length, w is the width, and h is the height. The volume of the original box is calculated:

$$V = 8 \times 14 \times 4 = 448 \text{ in}^3$$

The volume of the new box is calculated:

$$V = 16 \times 28 \times 8 = 3584 \text{ in}^3$$

The volume of the new box divided by the volume of the old box equals 8.

28. A: The notation i stands for an imaginary number. The value of i is equal to $\sqrt{-1}$. When performing calculations with imaginary numbers, treat i as a variable, and simplify when possible. Multiplying the binomials by the FOIL method produces $15 - 12i + 10i - 8i^2$. Combining like terms yields:

$$15 - 2i - 8i^2$$

Since:

$$i = \sqrt{-1}, i^2 = (\sqrt{-1})^2 = -1$$

Therefore, substitute -1 for i^2:

$$15 - 2i - 8(-1)$$

Simplifying results in:

$$15 - 2i + 8 \rightarrow 23 - 2i$$

29. C: The formula to find arc length is $s = \theta r$ where s is the arc length, θ is the radian measure of the central angle, and r is the radius of the circle. Substituting the given information produces:

$$3\pi \text{ cm} = \theta 12 \text{ cm}$$

Solving for θ yields $\theta = \frac{\pi}{4}$. To convert from radian to degrees, multiply the radian measure by:

$$\frac{180}{\pi} : \frac{\pi}{4} \times \frac{180}{\pi} = 45^o$$

30. A: The common denominator here will be 4x. Rewrite these fractions as

$$\frac{3}{x} + \frac{5u}{2x} - \frac{u}{4} = \frac{12}{4x} + \frac{10u}{4x} - \frac{ux}{4x} = \frac{12x + 10u - ux}{4x}$$

31. B: To simplify this inequality, subtract 3 from both sides to get $-\frac{1}{2}x \geq -1$. Then, multiply both sides by -2 (remembering this flips the direction of the inequality) to get $x \leq 2$.

32. D: This problem involves a composition function, where one function is plugged into the other function. In this case, the $f(x)$ function is plugged into the $g(x)$ function for each x-value. The composition equation becomes:

$$g(f(x)) = 2^3 - 3(2^2) - 2(2) + 6$$

Simplifying the equation gives the answer:

$$g(f(x)) = 8 - 3(4) - 2(2) + 6$$

$$8 - 12 - 4 + 6 = -2$$

33. D: A parabola of the form $y = \frac{1}{4f}x^2$ has a focus $(0, f)$. Because $y = -9x^2$, set $-9 = \frac{1}{4f}$. Solving this equation for f results in $f = -\frac{1}{36}$. Therefore, the coordinates of the focus are $\left(0, -\frac{1}{36}\right)$.

34. C: $51.93

List the givens.

$$Tax = 6.0\% = 0.06$$

$$Sale = 50\% = 0.5$$

$$Hat = \$32.99$$

$$Jersey = \$64.99$$

Calculate the sales prices.

$$Hat\ Sale = 0.5\ (32.99) = 16.495$$

$$Jersey\ Sale = 0.5\ (64.99) = 32.495$$

Total the sales prices.

$$Hat\ sale + jersey\ sale = 16.495 + 32.495 = 48.99$$

Calculate the tax and add it to the total sales prices.

$$Total\ after\ tax = 48.99 + (48.99\ x\ 0.06) = \$51.93$$

35. D: $0.45

List the givens.

$$Store\ coffee = \$1.23/lbs$$

$$Local\ roaster\ coffee = \$1.98/1.5\ lbs$$

Calculate the cost for 5 lbs of store brand.

$$\frac{\$1.23}{1\ lbs} \times 5\ lbs\ = \$6.15$$

Calculate the cost for 5 lbs of the local roaster.

$$\frac{\$1.98}{1.5\ lbs} \times 5\ lbs\ = \$6.60$$

Subtract to find the difference in price for 5 lbs.

$$\begin{array}{r} \$6.60 \\ -\ \$6.15 \\ \hline \$0.45 \end{array}$$

36. D: The shape of the scatterplot is a parabola (U-shaped). This eliminates Choices A (a linear equation that produces a straight line) and C (an exponential equation that produces a smooth curve upward or downward). The value of a for a quadratic function in standard form ($y = ax^2 + bx + c$) indicates whether the parabola opens up (U-shaped) or opens down (upside-down U). A negative value for a produces a parabola that opens down; therefore, Choice B can also be eliminated.

37. A: The volume of a cylinder is $\pi r^2 h$, and $\pi \times 5^2 \times 10$ is $250\ \pi\ in^3$. Choice B is not the correct answer because that is $5^2 \times 2\pi$. Choice C is not the correct answer since that is $5in \times 10\pi$. Choice D is not the correct answer because that is $10^2 \pi \times 2in$.

38. D: This system of equations involves one quadratic function and one linear function, as seen from the degree of each equation. One way to solve this is through substitution. Solving for y in the second equation yields $y = x + 2$. Plugging this equation in for the y of the quadratic equation yields:

$$x^2 - 2x + x + 2 = 8$$

Simplifying the equation, it becomes:

$$x^2 - x + 2 = 8$$

Setting this equal to zero and factoring, it becomes:

$$x^2 - x - 6 = 0 = (x - 3)(x + 2)$$

Solving these two factors for x gives the zeros $x = 3, -2$. To find the y-value for the point, each number can be plugged in to either original equation. Solving each one for y yields the points $(3, 5)$ and $(-2, 0)$.

39. B: For an ordered pair to be a solution to a system of inequalities, it must make a true statement for BOTH inequalities when substituting its values for x and y. Substituting $(-3, -2)$ into the inequalities produces $(-2) > 2(-3) - 3 \rightarrow -2 > -9$ and $(-2) < -4(-3) + 8 \rightarrow -2 < 20$. Both are true statements.

40.

If 60% of 50 workers are women, then there are 30 women working in the office. If half of them are wearing skirts, then that means 15 women wear skirts. Since none of the men wear skirts, this means there are 15 people wearing skirts.

41. A: Let the unknown score be x. The average will be:

$$\frac{5 \times 50 + 4 \times 70 + x}{10} = \frac{530 + x}{10} = 55$$

Multiply both sides by 10 to get $530 + x = 550$, or $x = 20$.

42. D: The addition rule is necessary to determine the probability because a 6 can be rolled on either roll of the die. The rule used is:

$$P(A \text{ or } B) = P(A) + P(B) - P(A \text{ and } B)$$

The probability of a 6 being individually rolled is $\frac{1}{6}$ and the probability of a 6 being rolled twice is:

$$\frac{1}{6} \times \frac{1}{6} = \frac{1}{36}$$

Therefore, the probability that a 6 is rolled at least once is:

$$\frac{1}{6} + \frac{1}{6} - \frac{1}{36} = \frac{11}{36}$$

43. A: Every 8 mL of medicine requires 5 mL. The 45 mL first needs to be split into portions of 8 mL. This results in $\frac{45}{8}$ portions. Each portion requires 5 mL. Therefore:

$$\frac{45}{8} \times 5 = \frac{45*5}{8} = \frac{225}{8} \text{ mL is necessary}$$

44. D: The midpoint formula should be used.

$$M = \left(\frac{x_1 + x_2}{2}, \frac{y_1 + y_2}{2}\right) = \left(\frac{-1 + 3}{2}, \frac{2 + (-6)}{2}\right) = (1, -2)$$

45. B: An equilateral triangle has three sides of equal length, so if the total perimeter is 18 feet, each side must be 6 feet long. A square with sides of 6 feet will have an area of $6^2 = 36$ square feet.

46. A: The formula for the volume of a sphere is $\frac{4}{3}\pi r^3$, and $\frac{4}{3} \times \pi \times 3^3$ is $36\,\pi$ in³. Choice *B* is not the correct answer because that is only 3^3. Choice *C* is not the correct answer because that is 3^2, and Choice *D* is not the correct answer because that is 36×2.

47.

$100 - 64$ is 36, and taking the square root of 36 is 6.

48. A: The formula for the area of the circle is πr^2 and 9 squared is 81. Choice *B* is not the correct answer because that is 2×9. Choice *C* is not the correct answer because that is 9×10. Choice *D* is not the correct answer because that is simply the value of the radius.

49. B: 13,078. The power of 10 by which a digit is multiplied corresponds with the number of zeros following the digit when expressing its value in standard form. Therefore:

$$(1 \times 10^4) + (3 \times 10^3) + (7 \times 10^1) + (8 \times 10^0)$$

$$10,000 + 3,000 + 70 + 8 = 13,078$$

50. A: Using the trigonometric identity $\tan(\theta) = \frac{\sin(\theta)}{\cos(\theta)}$, the expression becomes $\frac{\sin\theta}{\cos\theta}\cos\theta$. The factors that are the same on the top and bottom cancel out, leaving the simplified expression $\sin\theta$.

51. B: $\frac{11}{15}$. Fractions must have like denominators to be added. We are trying to add a fraction with a denominator of 3 to a fraction with a denominator of 5, so we have to convert both fractions to their respective equivalent fractions that have a common denominator. The common denominator is the least common multiple (LCM) of the two original denominators. In this case, the LCM is 15, so both fractions should be changed to equivalent fractions with a denominator of 15. To determine the numerator of the new fraction, the old numerator is multiplied by the same number by which the old denominator is multiplied to obtain the new denominator. For the fraction $\frac{2}{5}$, multiplying both the numerator and denominator by 3 produces $\frac{6}{15}$. When fractions have like denominators, they are added by adding the numerators and keeping the denominator the same:

$$\frac{5}{15} + \frac{6}{15} = \frac{11}{15}$$

52. B: $90° - 30° = 60°$. Choice *A* is not the correct answer because that is simply the original angle given. Choice *C* is not the correct answer since that is the angle you subtract from. Choice *D* is not the correct answer because that is $90° + 30°$.

53.

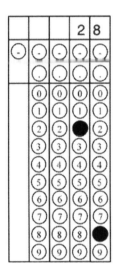

Each number in the sequence is adding one more than the difference between the previous two. For example:

$$10 - 6 = 4, 4 + 1 = 5$$

Therefore, the next number after 10 is $10 + 5 = 15$. Going forward, $21 - 15 = 6, 6 + 1 = 7$. The next number is $21 + 7 = 28$. Therefore, the difference between numbers is the set of whole numbers starting at 2: 2, 3, 4, 5, 6, 7….

54.

If a line is parallel to a side of a triangle and intersects the other two sides of the triangle, it separates the sides into corresponding segments of proportional lengths. To solve, set up a proportion:

$$\frac{AE}{AD} = \frac{AB}{AC} \rightarrow \frac{4}{5} = \frac{5}{x}$$

Cross multiplying yields:

$$4x = 25 \rightarrow x = 6.25$$

55.

		0	.	1	2

The fraction is converted so that the denominator is 100 by multiplying the numerator and denominator by 4, to get $\frac{3}{25} = \frac{12}{100}$. Dividing a number by 100 just moves the decimal point two places to the left, with a result of 0.12.

56.

				2	0

30% is $\frac{3}{10}$. The number itself must be $\frac{10}{3}$ of 6, or $\frac{10}{3} \times 6 = 10 \times 2 = 20$.

57.

			1	0

8 squared is 64, and 6 squared is 36. These should be added together to get $64 + 36 = 100$. Then, the last step is to find the square root of 100 which is 10.

58.

		1	1	2

Because the 68-degree angle and angle *b* sum to 180 degrees, the measurement of angle *b* is 112 degrees. Because of corresponding angles, angle *b* is equal to angle *f*. Therefore, angle *f* measures 112 degrees.

Practice Test #2

Reading Test

Passage #1

Questions 1–6 are based on the following passage:

When researchers and engineers undertake a large-scale scientific project, they may end up making discoveries and developing technologies that have far wider uses than originally intended. This is especially true in NASA, one of the most influential and innovative scientific organizations in America. NASA *spinoff technology* refers to innovations originally developed for NASA space projects that are now used in a wide range of different commercial fields. Many consumers are unaware that products they are buying are based on NASA research! Spinoff technology proves that it is worthwhile to invest in science research because it could enrich people's lives in unexpected ways.

The first spinoff technology worth mentioning is baby food. In space, where astronauts have limited access to fresh food and fewer options about their daily meals, malnutrition is a serious concern. Consequently, NASA researchers were looking for ways to enhance the nutritional value of astronauts' food. Scientists found that a certain type of algae could be added to food, improving the food's neurological benefits. When experts in the commercial food industry learned of this algae's potential to boost brain health, they were quick to begin their own research. The nutritional substance from algae then developed into a product called life's DHA, which can be found in over 90% of infant food sold in America.

Another intriguing example of a spinoff technology can be found in fashion. People who are always dropping their sunglasses may have invested in a pair of sunglasses with scratch resistant lenses—that is, it's impossible to scratch the glass, even if the glasses are dropped on an abrasive surface. This innovation is incredibly advantageous for people who are clumsy, but most shoppers don't know that this technology was originally developed by NASA. Scientists first created scratch resistant glass to help protect costly and crucial equipment from getting scratched in space, especially the helmet visors in space suits. However, sunglasses companies later realized that this technology could be profitable for their products, and they licensed the technology from NASA.

1. What is the main purpose of this article?
   a. To advise consumers to do more research before making a purchase
   b. To persuade readers to support NASA research
   c. To tell a narrative about the history of space technology
   d. To define and describe instances of spinoff technology

2. What is the organizational structure of this article?
   a. A general definition followed by more specific examples
   b. A general opinion followed by supporting arguments
   c. An important moment in history followed by chronological details
   d. A popular misconception followed by counterevidence

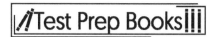

3. Why did NASA scientists research algae?
    a. They already knew algae was healthy for babies.
    b. They were interested in how to grow food in space.
    c. They were looking for ways to add health benefits to food.
    d. They hoped to use it to protect expensive research equipment.

4. What does the word "neurological" mean in the second paragraph?
    a. Related to the body
    b. Related to the brain
    c. Related to vitamins
    d. Related to technology

5. Why does the author mention space suit helmets?
    a. To give an example of astronaut fashion
    b. To explain where sunglasses got their shape
    c. To explain how astronauts protect their eyes
    d. To give an example of valuable space equipment

6. Which statement would the author probably NOT agree with?
    a. Consumers don't always know the history of the products they are buying.
    b. Sometimes new innovations have unexpected applications.
    c. It is difficult to make money from scientific research.
    d. Space equipment is often very expensive.

Passage #2

Questions 7–14 are based on the following passage:

People who argue that William Shakespeare is not responsible for the plays attributed to his name are known as anti-Stratfordians (from the name of Shakespeare's birthplace, Stratford-upon-Avon). The most common anti-Stratfordian claim is that William Shakespeare simply was not educated enough or from a high enough social class to have written plays overflowing with references to such a wide range of subjects like history, the classics, religion, and international culture. William Shakespeare was the son of a glove-maker, he only had a basic grade school education, and he never set foot outside of England—so how could he have produced plays of such sophistication and imagination? How could he have written in such detail about historical figures and events, or about different cultures and locations around Europe? According to anti-Stratfordians, the depth of knowledge contained in Shakespeare's plays suggests a well-traveled writer from a wealthy background with a university education, not a countryside writer like Shakespeare. But in fact, there is not much substance to such speculation, and most anti-Stratfordian arguments can be refuted with a little background about Shakespeare's time and upbringing.

First of all, those who doubt Shakespeare's authorship often point to his common birth and brief education as stumbling blocks to his writerly genius. Although it is true that Shakespeare did not come from a noble class, his father was a very *successful* glove-maker and his mother was from a very wealthy landowning family—so while Shakespeare may have had a country upbringing, he was certainly from a well-off family and would have been educated accordingly. Also, even though he did not attend university, grade school education in Shakespeare's time was actually

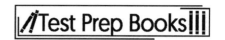

quite rigorous and exposed students to classic drama through writers like Seneca and Ovid. It is not unreasonable to believe that Shakespeare received a very solid foundation in poetry and literature from his early schooling.

Next, anti-Stratfordians tend to question how Shakespeare could write so extensively about countries and cultures he had never visited before (for instance, several of his most famous works like *Romeo and Juliet* and *The Merchant of Venice* were set in Italy, on the opposite side of Europe!). But again, this criticism does not hold up under scrutiny. For one thing, Shakespeare was living in London, a bustling metropolis of international trade, the most populous city in England, and a political and cultural hub of Europe. In the daily crowds of people, Shakespeare would certainly have been able to meet travelers from other countries and hear firsthand accounts of life in their home country. And, in addition to the influx of information from world travelers, this was also the age of the printing press, a jump in technology that made it possible to print and circulate books much more easily than in the past. This also allowed for a freer flow of information across different countries, allowing people to read about life and ideas from throughout Europe. One needn't travel the continent in order to learn and write about its culture.

7. What is the main purpose of this article?
    a. To explain two sides of an argument and allow readers to choose which side they agree with
    b. To encourage readers to be skeptical about the authorship of famous poems and plays
    c. To give historical background about an important literary figure
    d. To criticize a theory by presenting counterevidence

8. Which sentence contains the author's thesis?
    a. People who argue that William Shakespeare is not responsible for the plays attributed to his name are known as anti-Stratfordians.
    b. But in fact, there is not much substance to such speculation, and most anti-Stratfordian arguments can be refuted with a little background about Shakespeare's time and upbringing.
    c. It is not unreasonable to believe that Shakespeare received a very solid foundation in poetry and literature from his early schooling.
    d. Next, anti-Stratfordians tend to question how Shakespeare could write so extensively about countries and cultures he had never visited before.

9. In the first paragraph, "How could he have written in such detail about historical figures and events, or about different cultures and locations around Europe?" is an example of which of the following?
    a. Hyperbole
    b. Onomatopoeia
    c. Rhetorical question
    d. Appeal to authority

10. How does the author respond to the claim that Shakespeare was not well-educated because he did not attend university?
    a. By insisting upon Shakespeare's natural genius
    b. By explaining grade school curriculum in Shakespeare's time
    c. By comparing Shakespeare with other uneducated writers of his time
    d. By pointing out that Shakespeare's wealthy parents probably paid for private tutors

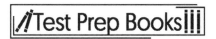

11. What does the word *bustling* in the third paragraph most nearly mean?
    a. Busy
    b. Foreign
    c. Expensive
    d. Undeveloped

12. What can be inferred from the article?
    a. Shakespeare's peers were jealous of his success and wanted to attack his reputation.
    b. Until recently, classic drama was only taught in universities.
    c. International travel was extremely rare in Shakespeare's time.
    d. In Shakespeare's time, glove-makers were not part of the upper class.

13. Why does the author mention *Romeo and Juliet*?
    a. It is Shakespeare's most famous play.
    b. It was inspired by Shakespeare's trip to Italy.
    c. It is an example of a play set outside of England.
    d. It was unpopular when Shakespeare first wrote it.

14. Which statement would the author probably agree with?
    a. It is possible to learn things from reading rather than from firsthand experience.
    b. If you want to be truly cultured, you need to travel the world
    c. People never become successful without a university education.
    d. All of the world's great art comes from Italy.

Passage #3

Questions 15–20 are based on the following excerpt from Dracula *by Bram Stoker, 1897:*

When I got on the coach the driver had not taken his seat, and I saw him talking with the landlady. They were evidently talking of me, for every now and then they looked at me, and some of the people who were sitting on the bench outside the door came and listened, and then looked at me, most of them pityingly. I could hear a lot of words often repeated, queer words, for there were many nationalities in the crowd; so I quietly got my polyglot dictionary from my bag and looked them out. I must say they weren't cheering to me, for amongst them were "Ordog"—Satan, "pokol"—hell, "stregoica"—witch, "vrolok" and "vlkoslak"—both of which mean the same thing, one being Slovak and the other Servian for something that is either were-wolf or vampire.

When we started, the crowd round the inn door, which had by this time swelled to a considerable size, all made the sign of the cross and pointed two fingers towards me. With some difficulty I got a fellow-passenger to tell me what they meant; he wouldn't answer at first, but on learning that I was English, he explained that it was a charm or guard against the evil eye. This was not very pleasant for me, just starting for an unknown place to meet an unknown man; but everyone seemed so kind-hearted, and so sorrowful, and so sympathetic that I couldn't but be touched. I shall never forget the last glimpse which I had of the inn-yard and its crowd of picturesque figures, all crossing themselves, as they stood round the wide archway, with its background of rich foliage of oleander and orange trees in green tubs clustered in the centre of the yard. Then our driver cracked his big whip over his four small horses, which ran abreast, and we set off on our journey.

I soon lost sight and recollection of ghostly fears in the beauty of the scene as we drove along, although had I known the language, or rather languages, which my fellow-passengers were speaking, I might not have been able to throw them off so easily. Before us lay a green sloping land full of forests and woods, with here and there steep hills, crowned with clumps of trees or with farmhouses, the blank gable end to the road. There was everywhere a bewildering mass of fruit blossom—apple, plum, pear, cherry; and as we drove by I could see the green grass under the trees spangled with the fallen petals. In and out amongst these green hills of what they call here the "Mittel Land" ran the road, losing itself as it swept round the grassy curve, or was shut out by the straggling ends of pine woods, which here and there ran down the hillsides like tongues of flame. The road was rugged, but still we seemed to fly over it with a feverish haste. I couldn't understand then what the haste meant, but the driver was evidently bent on losing no time in reaching Borgo Prund.

15. What type of narrator is found in this passage?
    a. First person
    b. Second person
    c. Third-person limited
    d. Third-person omniscient

16. Which of the following is true of the traveler?
    a. He wishes the driver would go faster.
    b. He's returning to the country of his birth.
    c. He has some familiarity with the local customs.
    d. He doesn't understand all of the languages being used.

17. How does the traveler's mood change between the second and third paragraphs?
    a. From relaxed to rushed
    b. From fearful to charmed
    c. From confused to enlightened
    d. From comfortable to exhausted

18. Who is the traveler going to meet?
    a. A kind landlady
    b. A distant relative
    c. A friendly villager
    d. A complete stranger

19. Based on the details in this passage, what can readers probably expect to happen in the story?
    a. The traveler will become a farmer.
    b. The traveler will arrive late at his destination.
    c. The traveler will soon encounter danger or evil.
    d. The traveler will have a pleasant journey and make many new friends.

20. Which sentence from the passage provides a clue for what will likely happen in the story?
    a. "I must say they weren't cheering to me, for amongst them were "Ordog"—Satan, "pokol"—hell, "stregoica"—witch, "vrolok" and "vlkoslak"—both of which mean the same thing, one being Slovak and the other Servian for something that is either were-wolf or vampire."
    b. "When I got on the coach the driver had not taken his seat, and I saw him talking with the landlady."
    c. "Then our driver cracked his big whip over his four small horses, which ran abreast, and we set off on our journey."
    d. "There was everywhere a bewildering mass of fruit blossom—apple, plum, pear, cherry; and as we drove by I could see the green grass under the trees spangled with the fallen petals."

Passage #4

Questions 21–26 are based on the following passage, which is an excerpt adapted from Robert Louis Stevenson's The Strange Case of Dr. Jekyll and Mr. Hyde.

"Did you ever come across a protégé of his—one Hyde?" He asked.

"Hyde?" repeated Lanyon. "No. Never heard of him. Since my time."

That was the amount of information that the lawyer carried back with him to the great, dark bed on which he tossed to and fro until the small hours of the morning began to grow large. It was a night of little ease to his toiling mind, toiling in mere darkness and besieged by questions.

Six o'clock struck on the bells of the church that was so conveniently near to Mr. Utterson's dwelling, and still he was digging at the problem. Hitherto it had touched him on the intellectual side alone; but; but now his imagination also was engaged, or rather enslaved; and as he lay and tossed in the gross darkness of the night in the curtained room, Mr. Enfield's tale went by before his mind in a scroll of lighted pictures. He would be aware of the great field of lamps in a nocturnal city; then of the figure of a man walking swiftly; then of a child running from the doctor's; and then these met, and that human Juggernaut trod the child down and passed on regardless of her screams. Or else he would see a room in a rich house, where his friend lay asleep, dreaming and smiling at his dreams; and then the door of that room would be opened, the curtains of the bed plucked apart, the sleeper recalled, and, lo! There would stand by his side a figure to whom power was given, and even at that dead hour he must rise and do its bidding. The figure in these two phrases haunted the lawyer all night; and if at anytime he dozed over, it was but to see it glide more stealthily through sleeping houses, or move the more swiftly, and still the more smoothly, even to dizziness, through wider labyrinths of lamplighted city, and at every street corner crush a child and leave her screaming. And still the figure had no face by which he might know it; even in his dreams it had no face, or one that baffled him and melted before his eyes; and thus there it was that there sprung up and grew apace in the lawyer's mind a singularly strong, almost an inordinate, curiosity to behold the features of the real Mr. Hyde. If he could but once set eyes on him, he thought the mystery would lighten and perhaps roll altogether away, as was the habit of mysterious things when well examined. He might see a reason for his friend's strange preference or bondage, and even for the startling clauses of the will. And at least it would be a face worth seeing: the face of a man who was without bowels of mercy: a face which had but to show itself to raise up, in the mind of the unimpressionable Enfield, a spirit of enduring hatred.

From that time forward, Mr. Utterson began to haunt the door in the by-street of shops. In the morning before office hours, at noon when business was plenty and time scarce, at night under the face of the full city moon, by all lights and at all hours of solitude or concourse, the lawyer was to be found on his chosen post.

"If he be Mr. Hyde," he had thought, "I should be Mr. Seek."

21. What is the purpose of the use of repetition in the following passage?
    It was a night of little ease to his toiling mind, toiling in mere darkness and besieged by questions.

    a. It serves as a demonstration of the mental state of Mr. Lanyon.
    b. It is reminiscent of the church bells that are mentioned in the story.
    c. It mimics Mr. Utterson's ambivalence.
    d. It emphasizes Mr. Utterson's anguish in failing to identify Hyde's whereabouts.

22. What is the setting of the story in this passage?
    a. In the city
    b. On the countryside
    c. In a jail
    d. In a mental health facility

23. What can one infer about the meaning of the word *Juggernaut* from the author's use of it in the passage?
    a. It is an apparition that appears at daybreak.
    b. It scares children.
    c. It is associated with space travel.
    d. Mr. Utterson finds it soothing.

24. What is the definition of the word *haunt* in the following passage?
    From that time forward, Mr. Utterson began to haunt the door in the by-street of shops. In the morning before office hours, at noon when business was plenty and time scarce, at night under the face of the full city moon, by all lights and at all hours of solitude or concourse, the lawyer was to be found on his chosen post.

    a. To levitate
    b. To constantly visit
    c. To terrorize
    d. To daunt

25. The phrase *labyrinths of lamplighted city* contains an example of what?
    a. Hyperbole
    b. Simile
    c. Juxtaposition
    d. Alliteration

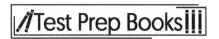

26. What can one reasonably conclude from the final comment of this passage?
    "If he be Mr. Hyde," he had thought, "I should be Mr. Seek."

    a. The speaker is considering a name change.
    b. The speaker is experiencing an identity crisis.
    c. The speaker has mistakenly been looking for the wrong person.
    d. The speaker intends to continue to look for Hyde.

Passage #5

Questions 27–32 are based on the following passage:

Dana Gioia argues in his article that poetry is dying, now little more than a limited art form confined to academic and college settings. Of course, poetry remains healthy in the academic setting, but the idea of poetry being limited to this academic subculture is a stretch. New technology and social networking alone have contributed to poets and other writers' work being shared across the world. YouTube has emerged to be a major asset to poets, allowing live performances to be streamed to billions of users. Even now, poetry continues to grow and voice topics that are relevant to the culture of our time. Poetry is not in the spotlight as it may have been in earlier times, but it's still a relevant art form that continues to expand in scope and appeal.

Furthermore, Gioia's argument does not account for live performances of poetry. Not everyone has taken a poetry class or enrolled in university—but most everyone is online. The Internet is a perfect launching point to get all creative work out there. An example of this was the performance of Buddy Wakefield's *Hurling Crowbirds at Mockingbars*. Wakefield is a well-known poet who has published several collections of contemporary poetry. One of my favorite works by Wakefield is *Crowbirds*, specifically his performance at New York University in 2009. Although his reading was a campus event, views of his performance online number in the thousands. His poetry attracted people outside of the university setting.

Naturally, the poem's popularity can be attributed both to Wakefield's performance and the quality of his writing. *Crowbirds* touches on themes of core human concepts such as faith, personal loss, and growth. These are not ideas that only poets or students of literature understand, but all human beings: "You acted like I was hurling crowbirds at mockingbars / and abandoned me for not making sense. / Evidently, I don't experience things as rationally as you do" (Wakefield 15-17). Wakefield weaves together a complex description of the perplexed and hurt emotions of the speaker undergoing a separation from a romantic interest. The line "You acted like I was hurling crowbirds at mockingbars" conjures up an image of someone confused, seemingly out of their mind . . . or in the case of the speaker, passionately trying to grasp at a relationship that is fading. The speaker is looking back and finding the words that described how he wasn't making sense. This poem is particularly human and gripping in its message, but the entire effect of the poem is enhanced through the physical performance.

At its core, poetry is about addressing issues/ideas in the world. Part of this is also addressing the perspectives that are exiguously considered. Although the platform may look different, poetry continues to have a steady audience due to the emotional connection the poet shares with the audience.

27. Which one of the following best explains how the passage is organized?
a. The author begins with a long definition of the main topic, and then proceeds to prove how that definition has changed over the course of modernity.
b. The author presents a puzzling phenomenon and uses the rest of the passage to showcase personal experiences in order to explain it.
c. The author contrasts two different viewpoints, then builds a case showing preference for one over the other.
d. The passage is an analysis of another theory that the author has no stake in.

28. The author of the passage would likely agree most with which of the following?
a. Buddy Wakefield is a genius and is considered at the forefront of modern poetry.
b. Poetry is not irrelevant; it is an art form that adapts to the changing time while containing its core elements.
c. Spoken word is the zenith of poetic forms and the premier style of poetry in this decade.
d. Poetry is on the verge of vanishing from our cultural consciousness.

29. Which one of the following words, if substituted for the word *exiguously* in the last paragraph, would LEAST change the meaning of the sentence?
a. Indolently
b. Inaudibly
c. Interminably
d. Infrequently

30. Which of the following is most closely analogous to the author's opinion of Buddy Wakefield's performance in relation to modern poetry?
a. Someone's refusal to accept that the Higgs Boson will validate the Standard Model.
b. An individual's belief that soccer will lose popularity within the next fifty years.
c. A professor's opinion that poetry contains the language of the heart, while fiction contains the language of the mind.
d. A student's insistence that psychoanalysis is a subset of modern psychology.

31. What is the primary purpose of the passage?
a. To educate readers on the development of poetry and describe the historical implications of poetry in media.
b. To disprove Dana Gioia's stance that poetry is becoming irrelevant and is only appreciated in academia.
c. To inform readers of the brilliance of Buddy Wakefield and to introduce them to other poets that have influenced contemporary poetry.
d. To prove that Gioia's article does have some truth to it and to shed light on its relevance to modern poetry.

32. What is the author's main reason for including the quote in the passage?

    a. The quote opens up opportunity to disprove Gioia's views.

    b. To demonstrate that people are still writing poetry even if the medium has changed in current times.

    c. To prove that poets still have an audience to write for even if the audience looks different than it did centuries ago.

    d. The quote illustrates the complex themes poets continue to address, which still draws listeners and appreciation.

Passage #6

Questions 33–38 are based on the following passage from The Curious Case of Benjamin Button *by F.S. Fitzgerald, 1922:*

As long ago as 1860 it was the proper thing to be born at home. At present, so I am told, the high gods of medicine have decreed that the first cries of the young shall be uttered upon the anesthetic air of a hospital, preferably a fashionable one. So young Mr. and Mrs. Roger Button were fifty years ahead of style when they decided, one day in the summer of 1860, that their first baby should be born in a hospital. Whether this anachronism had any bearing upon the astonishing history I am about to set down will never be known.

I shall tell you what occurred, and let you judge for yourself.

The Roger Buttons held an enviable position, both social and financial, in ante-bellum Baltimore. They were related to the This Family and the That Family, which, as every Southerner knew, entitled them to membership in that enormous peerage which largely populated the Confederacy. This was their first experience with the charming old custom of having babies— Mr. Button was naturally nervous. He hoped it would be a boy so that he could be sent to Yale College in Connecticut, at which institution Mr. Button himself had been known for four years by the somewhat obvious nickname of "Cuff."

On the September morning consecrated to the enormous event he arose nervously at six o'clock dressed himself, adjusted an impeccable stock, and hurried forth through the streets of Baltimore to the hospital, to determine whether the darkness of the night had borne in new life upon its bosom.

When he was approximately a hundred yards from the Maryland Private Hospital for Ladies and Gentlemen he saw Doctor Keene, the family physician, descending the front steps, rubbing his hands together with a washing movement—as all doctors are required to do by the unwritten ethics of their profession.

Mr. Roger Button, the president of Roger Button & Co., Wholesale Hardware, began to run toward Doctor Keene with much less dignity than was expected from a Southern gentleman of that picturesque period. "Doctor Keene!" he called. "Oh, Doctor Keene!"

The doctor heard him, faced around, and stood waiting, a curious expression settling on his harsh, medicinal face as Mr. Button drew near.

"What happened?" demanded Mr. Button, as he came up in a gasping rush. "What was it? How is she? A boy? Who is it? What—"

"Talk sense!" said Doctor Keene sharply. He appeared somewhat irritated.

"Is the child born?" begged Mr. Button.

Doctor Keene frowned. "Why, yes, I suppose so—after a fashion." Again he threw a curious glance at Mr. Button.

33. What major event is about to happen in this story?
    a. Mr. Button is about to go to a funeral.
    b. Mr. Button's wife is about to have a baby.
    c. Mr. Button is getting ready to go to the doctor's office.
    d. Mr. Button is about to go shopping for new clothes.

34. What kind of tone does the above passage have?
    a. Nervous and Excited
    b. Sad and Angry
    c. Shameful and Confused
    d. Grateful and Joyous

35. What is the meaning of the word "consecrated" in paragraph 4?
    a. Numbed
    b. Chained
    c. Dedicated
    d. Moved

36. What does the author mean to do by adding the following statement?
    . . . rubbing his hands together with a washing movement—as all doctors are required to do by the unwritten ethics of their profession.

    a. Suggesting that Mr. Button is tired of the doctor.
    b. Trying to explain the detail of the doctor's profession.
    c. Hinting to readers that the doctor is an unethical man.
    d. Giving readers a visual picture of what the doctor is doing.

37. Which of the following best describes the development of this passage?
    a. It starts in the middle of a narrative in order to transition smoothly to a conclusion.
    b. It is a chronological narrative from beginning to end.
    c. The sequence of events is backwards—we go from future events to past events.
    d. It introduces the setting of the story and its characters.

38. Which of the following is an example of an imperative sentence?
    a. "Oh, Doctor Keene!"
    b. "Talk sense!"
    c. "Is the child born?"
    d. "Why, yes, I suppose so—"

Passage #7

Questions 39–44 are based on the following excerpt from "The Story of An Hour" by Kate Chopin, 1894:

Knowing that Mrs. Mallard was afflicted with heart trouble, great care was taken to break to her as gently as possible the news of her husband's death.

It was her sister Josephine who told her, in broken sentences; veiled hints that revealed in half concealing. Her husband's friend Richards was there, too, near her. It was he who had been in the newspaper office when intelligence of the railroad disaster was received, with Brently Mallard's name leading the list of "killed." He had only taken the time to assure himself of its truth by a second telegram, and had hastened to forestall any less careful, less tender friend in bearing the sad message.

She did not hear the story as many women have heard the same, with a paralyzed inability to accept its significance. She wept at once, with sudden, wild abandonment, in her sister's arms. When the storm of grief had spent itself she went away to her room alone. She would have no one follow her.

There stood, facing the open window, a comfortable, roomy armchair. Into this she sank, pressed down by a physical exhaustion that haunted her body and seemed to reach into her soul.

She could see in the open square before her house the tops of trees that were all aquiver with the new spring life. The delicious breath of rain was in the air. In the street below a peddler was crying his wares. The notes of a distant song which some one was singing reached her faintly, and countless sparrows were twittering in the eaves.

There were patches of blue sky showing here and there through the clouds that had met and piled one above the other in the west facing her window.

She sat with her head thrown back upon the cushion of the chair, quite motionless, except when a sob came up into her throat and shook her, as a child who has cried itself to sleep continues to sob in its dreams.

She was young, with a fair, calm face, whose lines bespoke repression and even a certain strength. But now here was a dull stare in her eyes, whose gaze was fixed away off yonder on one of those patches of blue sky. It was not a glance of reflection, but rather indicated a suspension of intelligent thought.

There was something coming to her and she was waiting for it, fearfully. What was it? She did not know; it was too subtle and elusive to name. But she felt it, creeping out of the sky, reaching toward her through the sounds, the scents, and color that filled the air.

Now her bosom rose and fell tumultuously. She was beginning to recognize this thing that was approaching to possess her, and she was striving to beat it back with her will—as powerless as her two white slender hands would have been. When she abandoned herself a little whispered word escaped her slightly parted lips. She said it over and over under her breath: "free, free, free!" The vacant stare and the look of terror that had followed it went from her eyes. They stayed keen and bright. Her pulses beat fast, and the coursing blood warmed and relaxed every inch of her body.

She did not stop to ask if it were or were not a monstrous joy that held her. A clear and exalted perception enabled her to dismiss the suggestion as trivial. She knew that she would weep again when she saw the kind, tender hands folded in death; the face that had never looked save with love upon her, fixed and gray and dead. But she saw beyond that bitter moment a long procession of years to come that would belong to her absolutely. And she opened and spread her arms out to them in welcome.

39. What point of view is the above passage told in?
   a. First person
   b. Second person
   c. Third person omniscient
   d. Third person limited

40. What kind of irony are we presented with in this story?
   a. The way Mrs. Mallard reacted to her husband's death.
   b. The way in which Mr. Mallard died.
   c. The way in which the news of her husband's death was presented to Mrs. Mallard.
   d. The way in which nature is compared with death in the story.

41. What is the meaning of the word "elusive" in paragraph 9?
   a. Horrible
   b. Indefinable
   c. Quiet
   d. Joyful

42. What is the best summary of the passage above?
   a. Mr. Mallard, a soldier during World War I, is killed by the enemy and leaves his wife widowed.
   b. Mrs. Mallard understands the value of friendship when her friends show up for her after her husband's death.
   c. Mrs. Mallard combats mental illness daily and will perhaps be sent to a mental institution soon.
   d. Mrs. Mallard, a newly widowed woman, finds unexpected relief in her husband's death.

43. What is the tone of this story?
   a. Confused
   b. Joyful
   c. Depressive
   d. All of the above

44. What is the meaning of the word "tumultuously" in paragraph 10?
   a. Orderly
   b. Unashamedly
   c. Violently
   d. Calmly

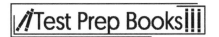

Passage #8

Questions 45–52 are based on the following passages:

Passage I

Lethal force, or deadly force, is defined as the physical means to cause death or serious harm to another individual. The law holds that lethal force is only accepted when you or another person are in immediate and unavoidable danger of death or severe bodily harm. For example, a person could be beating a weaker person in such a way that they are suffering severe enough trauma that could result in death or serious harm. This would be an instance where lethal force would be acceptable and possibly the only way to save that person from irrevocable damage.

Another example of when to use lethal force would be when someone enters your home with a deadly weapon. The intruder's presence and possession of the weapon indicate mal-intent and the ability to inflict death or severe injury to you and your loved ones. Again, lethal force can be used in this situation. Lethal force can also be applied to prevent the harm of another individual. If a woman is being brutally assaulted and is unable to fend off an attacker, lethal force can be used to defend her as a last-ditch effort. If she is in immediate jeopardy of rape, harm, and/or death, lethal force could be the only response that could effectively deter the assailant.

The key to understanding the concept of lethal force is the term *last resort*. Deadly force cannot be taken back; it should be used only to prevent severe harm or death. The law does distinguish whether the means of one's self-defense is fully warranted, or if the individual goes out of control in the process. If you continually attack the assailant after they are rendered incapacitated, this would be causing unnecessary harm, and the law can bring charges against you. Likewise, if you kill an attacker unnecessarily after defending yourself, you can be charged with murder. This would move lethal force beyond necessary defense, making it no longer a last resort but rather a use of excessive force.

Passage II

Assault is the unlawful attempt of one person to apply apprehension on another individual by an imminent threat or by initiating offensive contact. Assaults can vary, encompassing physical strikes, threatening body language, and even provocative language. In the case of the latter, even if a hand has not been laid, it is still considered an assault because of its threatening nature.

Let's look at an example: A homeowner is angered because his neighbor blows fallen leaves into his freshly mowed lawn. Irate, the homeowner gestures a fist to his fellow neighbor and threatens to bash his head in for littering on his lawn. The homeowner's physical motions and verbal threat heralds a physical threat against the other neighbor. These factors classify the homeowner's reaction as an assault. If the angry neighbor hits the threatening homeowner in retaliation, that would constitute an assault as well because he physically hit the homeowner.

Assault also centers on the involvement of weapons in a conflict. If someone fires a gun at another person, this could be interpreted as an assault unless the shooter acted in self-defense. If an individual drew a gun or a knife on someone with the intent to harm them, that would be considered assault. However, it's also considered an assault if someone simply aimed a weapon, loaded or not, at another person in a threatening manner.

45. What is the purpose of the second passage?
    a. To inform the reader about what assault is and how it is committed
    b. To inform the reader about how assault is a minor example of lethal force
    c. To disprove the previous passage concerning lethal force
    d. The author is recounting an incident in which they were assaulted

46. Which of the following situations, according to the passages, would not constitute an illegal use of lethal force?
    a. A disgruntled cashier yells obscenities at a customer.
    b. A thief is seen running away with stolen cash.
    c. A man is attacked in an alley by another man with a knife.
    d. A woman punches another woman in a bar.

47. Given the information in the passages, which of the following must be true about assault?
    a. Assault charges are more severe than unnecessary use of force charges.
    b. There are various forms of assault.
    c. Smaller, weaker people cannot commit assaults.
    d. Assault is justified only as a last resort.

48. Which of the following, if true, would most seriously undermine the explanation proposed by the author of Passage I in the third paragraph?
    a. An instance of lethal force in self-defense is not absolutely absolved from blame. The law considers the necessary use of force at the time it is committed.
    b. An individual who uses lethal force under necessary defense is in direct compliance of the law under most circumstances.
    c. Lethal force in self-defense should be forgiven in all cases for the peace of mind of the primary victim.
    d. The use of lethal force is not evaluated on the intent of the user but rather the severity of the primary attack that warranted self-defense.

49. Based on the passages, what can be inferred about the relationship between assault and lethal force?
    a. An act of lethal force always leads to a type of assault.
    b. An assault will result in someone using lethal force.
    c. An assault with deadly intent can lead to an individual using lethal force to preserve their well-being.
    d. If someone uses self-defense in a conflict, it is called deadly force; if actions or threats are intended, it is called assault.

50. Which of the following best describes the way the passages are structured?
    a. Both passages open by defining a legal concept and then continue to describe situations that further explain the concept.
    b. Both passages begin with situations, introduce accepted definitions, and then cite legal ramifications.
    c. Passage I presents a long definition while the Passage II begins by showing an example of assault.
    d. Both cite specific legal doctrines, then proceed to explain the rulings.

51. What can be inferred about the role of intent in lethal force and assault?
    a. Intent is irrelevant. The law does not take intent into account.
    b. Intent is vital for determining the lawfulness of using lethal force.
    c. Intent is very important for determining both lethal force and assault; intent is examined in both parties and helps determine the severity of the issue.
    d. The intent of the assailant is the main focus for determining legal ramifications; it is used to determine if the defender was justified in using force to respond.

52. The author uses the example in the second paragraph of Passage II in order to do what?
    a. To demonstrate two different types of assault by showing how each specifically relates to the other
    b. To demonstrate a single example of two different types of assault, then adding in the third type of assault in the example's conclusion
    c. To prove that the definition of lethal force is altered when the victim in question is a homeowner and his property is threatened
    d. To suggest that verbal assault can be an exaggerated crime by the law and does not necessarily lead to physical violence

Writing and Language Test

Read the essay entitled "Education is Essential to Civilization" and answer Questions 1–15.

Early in my career, (1) <u>a master's teacher shared this thought with me "Education is the last bastion of civility."</u> While I did not completely understand the scope of those words at the time, I have since come to realize the depth, breadth, truth, and significance of what he said. (2) <u>Education provides</u> society with a vehicle for (3) <u>raising it's children to be</u> civil, decent, human beings with something valuable to contribute to the world. It is really what makes us human and what (4) <u>distinguishes</u> us as <u>civilised</u> <u>creatures.</u>

Being "civilized" humans means being "whole" humans. Education must address the mind, body, and soul of students. (5) <u>It would be detrimental to society, only meeting the needs of the mind, if our schools were myopic in their focus.</u> As humans, we are multi-dimensional, multi-faceted beings who need more than head knowledge to survive. (6) <u>The human heart and psyche have to be fed in order for the mind to develop properly, and the body must be maintained and exercised to help fuel the working of the brain. Education is a basic human right, and it allows us to sustain a democratic society in which participation is fundamental to its success. It should inspire students to seek better solutions to world problems and to dream of a more equitable society.</u> Education should never discriminate on any basis, and it should create individuals who are self-sufficient, patriotic, and tolerant of (7) <u>others' ideas.</u>

(8) <u>All children can learn. Although not all children learn in the same manner.</u> All children learn best, however, when their basic physical needs are met and they feel safe, secure, and loved. Students are much more responsive to a teacher who values them and shows them respect as individual people. Teachers must model at all times the way they expect students to treat them and their peers. If teachers set high expectations for (9) <u>there students</u>, the students will rise to that high level. Teachers must make the well-being of students their primary focus and must not be afraid to let students learn from their own mistakes.

In the modern age of technology, a teacher's focus is no longer the "what" of the content, (10) <u>but more importantly, the 'why.'</u> Students are bombarded with information and have access to ANY information they need right at their fingertips. Teachers have to work harder than ever before to help students identify salient information (11) <u>so to think critically</u> about the information they encounter. Students have to (12) <u>read between the lines, identify bias, and determine</u> who they can trust in the milieu of ads, data, and texts presented to them.

Schools must work in consort with families in this important mission. While children spend most of their time in school, they are dramatically and indelibly shaped (13) <u>with the influences</u> of their family and culture. Teachers must not only respect this fact, (14) <u>but must strive</u> to include parents in the education of their children and must work to keep parents informed of progress and problems. Communication between classroom and home is essential for a child's success.

Humans have always aspired to be more, do more, and to better ourselves and our communities. This is where education lies, right at the heart of humanity's desire to be all that we can be. Education helps us strive for higher goals and better treatment of ourselves and others. I shudder to think what would become of us if education ceased to be the "last bastion of civility." (15) <u>We must be unapologetic about expecting excellence from our students? Our very existence depends upon it.</u>

1. What edit is needed to correct sentence 1 (reproduced below)?

Early in my career, (1) <u>a master's teacher shared this thought with me "Education is the last bastion of civility."</u>

a. NO CHANGE
b. a master's teacher shared this thought with me: "Education is the last bastion of civility."
c. a master's teacher shared this thought with me: "Education is the last bastion of civility".
d. a master's teacher shared this thought with me. "Education is the last bastion of civility."

2. What edit is needed to correct sentence 2 (reproduced below)?

(2) <u>Education provides</u> society with a vehicle for raising it's children to be civil, decent, human beings with something valuable to contribute to the world.

a. NO CHANGE
b. Education provide
c. Education will provide
d. Education providing

3. What edit is needed to correct sentence 3 (reproduced below)?

Education provides society with a vehicle for (3) <u>raising it's children to be</u> civil, decent, human beings with something valuable to contribute to the world.

a. NO CHANGE
b. raises its children to be
c. raising its' children to be
d. raising its children to be

4. Which of these, if any, is misspelled?
a. None of these are misspelled.
b. distinguishes
c. civilised
d. creatures

5. What edit is needed to correct sentence 5 (reproduced below)?

(5) <u>It would be detrimental to society, only meeting the needs of the mind, if our schools were myopic in their focus.</u>

a. NO CHANGE
b. It would be detrimental to society if our schools were myopic in their focus, only meeting the needs of the mind.
c. Only meeting the needs of our mind, our schools were myopic in their focus, detrimental to society.
d. Myopic is the focus of our schools, being detrimental to society for only meeting the needs of the mind.

6. Which of these sentences, if any, should begin a new paragraph?
a. NO CHANGE
b. The human heart and psyche have to be fed in order for the mind to develop properly, and the body must be maintained and exercised to help fuel the working of the brain.
c. Education is a basic human right, and it allows us to sustain a democratic society in which participation is fundamental to its success.
d. It should inspire students to seek better solutions to world problems and to dream of a more equitable society.

7. What edit is needed to correct sentence 7 (reproduced below)?

Education should never discriminate on any basis, and it should create individuals who are self-sufficient, patriotic, and tolerant of (7) <u>others' ideas.</u>

a. NO CHANGE
b. other's ideas
c. others ideas
d. others's ideas

8. What edit is needed to correct sentence 8 (reproduced below)?

(8) <u>All children can learn. Although not all children learn in the same manner.</u>

a. NO CHANGE
b. All children can learn although not all children learn in the same manner.
c. All children can learn although, not all children learn in the same manner.
d. All children can learn, although not all children learn in the same manner.

9. What edit is needed to correct sentence 9 (reproduced below)?

If teachers set high expectations for (9) <u>there students</u>, the students will rise to that high level.

a. NO CHANGE
b. they're students
c. their students
d. thare students

10. What edit is needed to correct sentence 10 (reproduced below)?

In the modern age of technology, a teacher's focus is no longer the "what" of the content, (10) <u>but more importantly, the 'why.'</u>

a. NO CHANGE
b. but more importantly, the "why."
c. but more importantly, the 'why'.
d. but more importantly, the "why".

11. What edit is needed to correct sentence 11 (reproduced below)?

Teachers have to work harder than ever before to help students identify salient information (11) <u>so to think critically</u> about the information they encounter.

a. NO CHANGE
b. and to think critically
c. but to think critically
d. nor to think critically

12. What edit is needed to correct sentence 12 (reproduced below)?

Students have to (12) <u>read between the lines, identify bias, and determine</u> who they can trust in the milieu of ads, data, and texts presented to them.

a. NO CHANGE
b. read between the lines, identify bias, and determining
c. read between the lines, identifying bias, and determining
d. reads between the lines, identifies bias, and determines

13. What edit is needed to correct sentence 13 (reproduced below)?

While children spend most of their time in school, they are dramatically and indelibly shaped (13) <u>with the influences</u> of their family and culture.

a. NO CHANGE
b. for the influences
c. to the influences
d. by the influences

14. What edit is needed to correct sentence 14 (reproduced below)?

Teachers must not only respect this fact, (14) <u>but must strive</u> to include parents in the education of their children and must work to keep parents informed of progress and problems.

a. NO CHANGE
b. but to strive
c. but striving
d. but strived

15. What edit is needed to correct sentence 15 (reproduced below)?

(15) <u>We must be unapologetic about expecting excellence from our students? Our very existence depends upon it.</u>

a. NO CHANGE
b. We must be unapologetic about expecting excellence from our students, our very existence depends upon it.
c. We must be unapologetic about expecting excellence from our students—our very existence depends upon it.
d. We must be unapologetic about expecting excellence from our students our very existence depends upon it.

Questions 16–24 are based on the following passage about Frankenstein *by Mary Shelley:*

(16) <u>One of the icon's of romantic and science fiction literature</u> remains Mary Shelley's classic, *Frankenstein, or The Modern Prometheus*. Schools throughout the world still teach the book in literature and philosophy courses. Scientific communities also engage in discussion on the novel. But why? Besides the novel's engaging (17) <u>writing style the story's central theme</u> remains highly relevant in a world of constant discovery and moral dilemmas. Central to the core narrative is the (18) <u>struggle between enlightenment and the cost of overusing power.</u>

The subtitle, *The Modern Prometheus*, encapsulates the inner theme of the story more than the main title of *Frankenstein*. As with many romantic writers, Shelley invokes the classical myths and (19) <u>symbolism of Ancient Greece and Rome to high light core ideas.</u> Looking deeper into the myth of Prometheus sheds light not only on the character of Frankenstein (20) <u>but also poses a psychological dilemma to the audience.</u> Prometheus is the titan who gave fire to mankind. (21) <u>However, more than just fire he gave people knowledge and power.</u> The power of fire advanced civilization. Yet, for giving fire to man, Prometheus is (22) <u>punished by the gods</u>

bound to a rock and tormented for his act. This is clearly a parallel to Frankenstein—he is the modern Prometheus.

Frankenstein's quest for knowledge becomes an obsession. It leads him to literally create new life, breaking the bounds of conceivable science to illustrate that man can create life out of nothing. (23) Yet he ultimately faltered as a creator, abandoning his progeny in horror of what he created. Frankenstein then suffers his creature's wrath, (24) the result of his pride, obsession for power and lack of responsibility.

Shelley isn't condemning scientific achievement. Rather, her writing reflects that science and discovery are good things, but, like all power, it must be used wisely. The text alludes to the message that one must have reverence for nature and be mindful of the potential consequences. Frankenstein did not take responsibility or even consider how his actions would affect others. His scientific brilliance ultimately led to suffering.

16. Which of the following would be the best choice for this sentence (reproduced below)?

(16) One of the icon's of romantic and science fiction literature remains Mary Shelley's classic, Frankenstein, or The Modern Prometheus.

a. NO CHANGE
b. One of the icons of romantic and science fiction literature
c. One of the icon's of romantic, and science fiction literature,
d. The icon of romantic and science fiction literature

17. Which of the following would be the best choice for this sentence (reproduced below)?

Besides the novel's engaging (17) writing style the story's central theme remains highly relevant in a world of constant discovery and moral dilemmas.

a. NO CHANGE
b. writing style the central theme of the story
c. writing style, the story's central theme
d. the story's central theme's writing style

18. Which of the following would be the best choice for this sentence (reproduced below)?

Central to the core narrative is the (18) struggle between enlightenment and the cost of overusing power.

a. NO CHANGE
b. struggle between enlighten and the cost of overusing power.
c. struggle between enlightenment's cost of overusing power.
d. struggle between enlightening and the cost of overusing power.

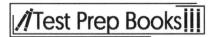

19. Which of the following would be the best choice for this sentence (reproduced below)?

> As with many romantic writers, Shelley invokes the classical myths and (19) symbolism of Ancient Greece and Rome to high light core ideas.

a. NO CHANGE
b. symbolism of Ancient Greece and Rome to highlight core ideas.
c. symbolism of ancient Greece and Rome to highlight core ideas.
d. symbolism of Ancient Greece and Rome highlighting core ideas.

20. Which of the following would be the best choice for this sentence (reproduced below)?

> Looking deeper into the myth of Prometheus sheds light not only on the character of Frankenstein (20) but also poses a psychological dilemma to the audience.

a. NO CHANGE
b. but also poses a psychological dilemma with the audience.
c. but also poses a psychological dilemma for the audience.
d. but also poses a psychological dilemma there before the audience.

21. Which of the following would be the best choice for this sentence (reproduced below)?

> (21) However, more than just fire he gave people knowledge and power.

a. NO CHANGE
b. However, more than just fire he gave people, knowledge, and power.
c. However, more than just fire, he gave people knowledge and power.
d. Besides actual fire, Prometheus gave people knowledge and power.

22. Which of the following would be the best choice for this sentence (reproduced below)?

> Yet, for giving fire to man, Prometheus is (22) punished by the gods bound to a rock and tormented for his act.

a. NO CHANGE
b. punished by the gods, bound to a rock and tormented for his act.
c. bound to a rock and tormented as punishment by the gods.
d. punished for his act by being bound to a rock and tormented as punishment from the gods.

23. Which of the following would be the best choice for this sentence (reproduced below)?

(23) <u>Yet he ultimately faltered as a creator,</u> abandoning his progeny in horror of what he created.

a. NO CHANGE
b. Yet, he ultimately falters as a creator by
c. Yet, he ultimately faltered as a creator,
d. Yet he ultimately falters as a creator by

24. Which of the following would be the best choice for this sentence (reproduced below)?
Frankenstein then suffers his creature's wrath, (24) <u>the result of his pride, obsession for power and lack of responsibility.</u>

a. NO CHANGE
b. the result of his pride, obsession for power and lacking of responsibility.
c. the result of his pride, obsession for power, and lack of responsibility.
d. the result of his pride and also his obsession for power and lack of responsibility.

Questions 25–33 are based on the following passage:

The power of legends continues to enthrall our imagination, provoking us both to wonder and explore. (25) <u>Who doesnt love a good legend?</u> Some say legends never (26) <u>die and this is certainly the case</u> for the most legendary creature of all, Bigfoot. To this day, people still claim sightings of the illusive cryptid. Many think of Bigfoot as America's monster, yet many nations have legends of a similar creature. In my own research I have found that Australia has the Yowie, China has the Yerin, and Russia has the Almas. (27) <u>Their all over the world, the</u> bigfoots and the legends tied to them. Does this mean they could exist?

There are many things to consider when addressing (28) <u>this question but the chief factor</u> is whether there is credible evidence. (29) <u>For science to formally recognize that such a species exists, there needs to be physical proof.</u> While people have found supposed footprints and even (30) <u>captured photos and film of the creature, this validity of such evidence is up for debate</u>. There is room for uncertainty. Most visual evidence is out of focus, thus (31) <u>there is often skepticism whether such images are real.</u> Some researchers have even claimed to have hair and blood samples, but still there is doubt in the scientific community. The reason is simple: there needs to be a body or living specimen found and actively studied in order to prove the Bigfoots' existence.

Yet, one cannot ignore the fact that (32) <u>hundreds of witnesses continuing to describe a creature</u> with uniform features all over the world. These bigfoot sightings aren't a modern occurrence either. Ancient civilizations have reported (33) <u>seeing Bigfoot as well including Native Americans.</u> It is from Native Americans that we gained the popular term Sasquatch, which is the primary name for the North American bigfoot. How does their testimony factor in? If indigenous people saw these animals, could they not have existed at some point? After all, when Europeans first arrived in Africa, they disbelieved the native accounts of the gorilla. But sure enough, Europeans eventually found gorillas and collected a body.

25. Which of the following would be the best choice for this sentence (reproduced below)?

(25) <u>Who doesnt love a good legend?</u>

a. NO CHANGE
b. Who does not love a good legend?
c. A good legend, who doesn't love one?
d. Who doesn't love a good legend?

26. Which of the following would be the best choice for this sentence (reproduced below)?

Some say legends never (26) <u>die and this is certainly the case</u> for the most legendary creature of all, Bigfoot.

a. NO CHANGE
b. die, and this is certainly the case
c. die; this is certainly the case
d. die. This is certainly the case

27. Which of the following would be the best choice for this sentence (reproduced below)?

(27) <u>Their all over the world, the</u> bigfoots and the legends tied to them.

a. NO CHANGE
b. There all over the world, the
c. They're all over the world, the
d. All over the world they are, the

28. Which of the following would be the best choice for this sentence (reproduced below)?

There are many things to consider when addressing (28) <u>this question but the chief factor</u> is whether there is credible evidence.

a. NO CHANGE
b. this question, but the chief factor
c. this question however the chief factor
d. this question; but the chief factor

29. Which of the following would be the best choice for this sentence (reproduced below)?

(29) <u>For science to formally recognize that such a species exists, there needs to be physical proof.</u>

a. NO CHANGE
b. Physical proof are needed in order for science to formally recognize that such a species exists.
c. For science to formally recognize that such a species exists there needs to be physical proof.
d. For science, to formally recognize that such a species exists, there needs to be physical proof.

30. Which of the following would be the best choice for this sentence (reproduced below)?

While people have found supposed footprints and even (30) captured photos and film of the creature, this validity of such evidence is up for debate.

a. NO CHANGE
b. captured photos and film of the creature. This validity of such evidence is up for debate.
c. captured photos and film of the creature, the validities of such evidence is up for debate.
d. captured photos and film of the creature, the validity of such evidence is up for debate.

31. Which of the following would be the best choice for this sentence (reproduced below)?

Most visual evidence is out of focus, thus (31) there is often skepticism whether such images are real.

a. NO CHANGE
b. often skepticism whether such images are real.
c. there is often skepticism, whether such images are real.
d. there is often skepticism weather such images are real.

32. Which of the following would be the best choice for this sentence (reproduced below)?

Yet, one cannot ignore the fact that (32) hundreds of witnesses continuing to describe a creature with uniform features all over the world.

a. NO CHANGE
b. hundreds of witnesses continuing to describing a creature
c. hundreds of witnesses continue to describe a creature
d. hundreds of the witnesses continue to described a creature

33. Which of the following would be the best choice for this sentence (reproduced below)?

Ancient civilizations have reported (33) seeing Bigfoot as well including Native Americans.

a. NO CHANGE
b. seeing Bigfoot, Native Americans as well.
c. seeing Bigfoot also the Native Americans.
d. seeing Bigfoot, including Native Americans.

Questions 34–40 are based on the following passage:

I have to admit that when my father bought a recreational vehicle (RV), I thought he was making a huge mistake. I didn't really know anything about RVs, but I knew that my dad was as big a "city slicker" as there was. (34) In fact, I even thought he might have gone a little bit crazy. On trips to the beach, he preferred to swim at the pool, and whenever he went hiking, he avoided touching any plants for fear that they might be poison ivy. Why would this man, with an almost irrational fear of the outdoors, want a 40-foot camping behemoth?

(35) The RV was a great purchase for our family and brought us all closer together. Every morning (36) we would wake up, eat breakfast, and broke camp. We laughed at our own

comical attempts to back The Beast into spaces that seemed impossibly small. (37) <u>We rejoiced as "hackers."</u> When things inevitably went wrong and we couldn't solve the problems on our own, we discovered the incredible helpfulness and friendliness of the RV community. (38) <u>We even made some new friends in the process.</u>

(39) <u>Above all, it allowed us to share adventures. While travelling across America</u>, which we could not have experienced in cars and hotels. Enjoying a campfire on a chilly summer evening with the mountains of Glacier National Park in the background or waking up early in the morning to see the sun rising over the distant spires of Arches National Park are memories that will always stay with me and our entire family. (40) <u>Those are also memories that my siblings and me</u> have now shared with our own children.

34. Which of the following would be the best choice for this sentence (reproduced below)?

(34) <u>In fact, I even thought he might have gone a little bit crazy.</u>

a. NO CHANGE
b. Move the sentence so that it comes before the preceding sentence.
c. Move the sentence to the end of the first paragraph.
d. Omit the sentence.

35. In context, which is the best version of the underlined portion of this sentence (reproduced below)?

(35) <u>The RV</u> was a great purchase for our family and brought us all closer together.

a. NO CHANGE
b. Not surprisingly, the RV
c. Furthermore, the RV
d. As it turns out, the RV

36. Which is the best version of the underlined portion of this sentence (reproduced below)?

Every morning (36) <u>we would wake up, eat breakfast, and broke camp.</u>

a. NO CHANGE
b. we would wake up, eat breakfast, and break camp.
c. would we wake up, eat breakfast, and break camp?
d. we are waking up, eating breakfast, and breaking camp.

37. Which is the best version of the underlined portion of this sentence (reproduced below)?

(37) <u>We rejoiced as "hackers."</u>

a. NO CHANGE
b. To a nagging problem of technology, we rejoiced as "hackers."
c. We rejoiced when we figured out how to "hack" a solution to a nagging technological problem.
d. To "hack" our way to a solution, we had to rejoice.

38. Which is the best version of the underlined portion of this sentence (reproduced below)?

(38) <u>We even made some new friends in the process.</u>

a. NO CHANGE
b. In the process was the friends we were making.
c. We are even making some new friends in the process.
d. We will make new friends in the process.

39. Which is the best version of the underlined portion of this sentence (reproduced below)?

(39) <u>Above all, it allowed us to share adventures. While travelling across America,</u> which we could not have experienced in cars and hotels.

a. NO CHANGE
b. Above all, it allowed us to share adventures while traveling across America
c. Above all, it allowed us to share adventures; while traveling across America
d. Above all, it allowed us to share adventures—while traveling across America

40. Which is the best version of the underlined portion of this sentence (reproduced below)?

(40) <u>Those are also memories that my siblings and me</u> have now shared with our own children.

a. NO CHANGE
b. Those are also memories that me and my siblings
c. Those are also memories that my siblings and I
d. Those are also memories that I and my siblings

Questions 41–45 are based on the following passage:

> We live in a savage world; that's just a simple fact. It is a time of violence, when the need for self-defense is imperative. (41) <u>Martial arts, like Ju-jitsu, still play a vital role in ones survival.</u> (42) <u>Ju-jitsu, however doesn't justify kicking people around</u>, even when being harassed or attacked. Today, laws prohibit the (43) <u>use of unnecessary force in self-defense; these serve to eliminate</u> beating someone to a pulp once they have been neutralized. Such laws are needed. Apart from being unnecessary to continually strike a person when (44) <u>their down, its immoral</u>. Such over-aggressive retaliation turns the innocent into the aggressor. Ju-jitsu provides a way for defending oneself while maintaining the philosophy of restraint and self-discipline. (45) <u>Integrated into its core philosophy,</u> Ju-jitsu tempers the potential to do great physical harm with respect for that power and for life.

41. Which of the following would be the best choice for this sentence (reproduced below)?

(41) <u>Martial arts, like Ju-jitsu, still play a vital role in ones survival.</u>

a. NO CHANGE
b. Martial arts, like Ju-jitsu, still play a vital role in one's survival.
c. Martial arts, like Ju-jitsu still play a vital role in ones survival.
d. Martial arts, like Ju-jitsu, still plays a vital role in one's survival.

42. Which of the following would be the best choice for this sentence (reproduced below)?

(42) <u>Ju-jitsu, however doesn't justify kicking people around,</u> even when being harassed or attacked.

a. NO CHANGE
b. Ju-jitsu, however, isn't justified by kicking people around,
c. However, Ju-jitsu doesn't justify kicking people around,
d. Ju-jitsu however doesn't justify kicking people around,

43. Which of the following would be the best choice for this sentence (reproduced below)?

Today, laws prohibit the (43) <u>use of unnecessary force in self-defense; these serve to eliminate</u> beating someone to a pulp once they have been neutralized.

a. NO CHANGE
b. use of unnecessary force in self-defense serving to eliminate
c. use of unnecessary force, in self-defense, these serve to eliminate
d. use of unnecessary force. In self-defense, these serve to eliminate

44. Which of the following would be the best choice for this sentence (reproduced below)?

Apart from being unnecessary to continually strike a person when (44) <u>their down, its immoral.</u>

a. NO CHANGE
b. their down, it's immoral.
c. they're down, its immoral.
d. they're down, it's immoral.

45. Which of the following would be the best choice for this sentence (reproduced below)?

(45) <u>Integrated into its core philosophy,</u> Ju-jitsu tempers the potential to do great physical harm with respect for that power, and for life.

a. NO CHANGE
b. Integrated into its core philosophy
c. Integrated into it's core philosophy
d. Integrated into its' core philosophy,

Math Test

1. If a car can travel 300 miles in 4 hours, how far can it go in an hour and a half?
a. 100 miles
b. 112.5 miles
c. 135.5 miles
d. 150 miles

2. At the store, Jan spends $90 on apples and oranges. Apples cost $1 each and oranges cost $2 each. If Jan buys the same number of apples as oranges, how many oranges did she buy?

3. Where does the point (–3, –4) lie in relation to the circle with the equation $(x)^2 + (y)^2 = 25$?
a. Inside of the circle.
b. Outside of the circle.
c. On the circle.
d. There is not enough information to tell.

4. A train traveling 50 miles per hour takes a trip lasting 3 hours. If a map has a scale of 1 inch per 10 miles, how many inches apart are the train's starting point and ending point on the map if it travelled in a straight line?

5. A traveler takes an hour to drive to a museum, spends 3 hours and 30 minutes there, and takes half an hour to drive home. What percentage of his or her time was spent driving?

    a. 15%
    b. 30%
    c. 40%
    d. 60%

6. A truck is carrying three cylindrical barrels. Their bases have a diameter of 2 feet, and they have a height of 3 feet. What is the total volume of the three barrels in cubic feet?

    a. 3π
    b. 9π
    c. 12π
    d. 15π

7. Greg buys a $10 lunch with 5% sales tax. He leaves a $2 tip after paying his bill. How much money does he spend?

    a. $12.50
    b. $12
    c. $13
    d. $13.25

8. Marty wishes to save $150 over a 4-day period. How much must Marty save each day on average?

    a. $37.50
    b. $35
    c. $45.50
    d. $41

9. Bernard can make $80 per day. If he needs to make $300 and only works full days, how many days will this take?

10. A couple buys a house for $150,000. They sell it for $165,000. By what percentage did the house's value increase?
    a. 10%
    b. 13%
    c. 15%
    d. 17%

11. A school has 15 teachers and 20 teaching assistants. They have 200 students. What is the ratio of faculty to students?
    a. 3:20
    b. 4:17
    c. 5:54
    d. 7:40

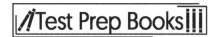
12. A map has a scale of 1 inch per 5 miles. A car can travel 60 miles per hour. If the distance from the start to the destination is 3 inches on the map, how many minutes will it take the car to make the trip?

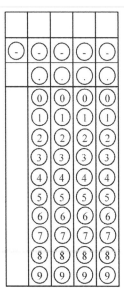

13. Taylor works two jobs. The first pays $20,000 per year. The second pays $10,000 per year. She donates 15% of her income to charity. How much does she donate each year?
    a. $4500
    b. $5000
    c. $5500
    d. $6000

14. Suppose an investor deposits $1,200 into a bank account that accrues 1 percent interest per month. Assuming x represents the number of months since the deposit and y represents the money in the account, which of the following exponential functions models the scenario?
    a. $y = (0.01)(1200^x)$
    b. $y = (1200)(0.01^x)$
    c. $y = (1.01)(1200^x)$
    d. $y = (1200)(1.01^x)$

15. Suppose the function $y = \frac{1}{8}x^3 + 2x - 21$ approximates the population of a given city between the years 1900 and 2000 with x representing the year ($1900 = 0$) and y representing the population (in 1000s). Which of the following domains are relevant for the scenario?
    a. $(-\infty, \infty)$
    b. [1900, 2000]
    c. [0, 100]
    d. [0, 0]

16. If Sarah reads at an average rate of 21 pages in four nights, how long will it take her to read 140 pages?

    a. 6 nights

    b. 26 nights

    c. 8 nights

    d. 27 nights

17. Mom's car drove 72 miles in 90 minutes. There are 5280 feet per mile. How fast did she drive in feet per second?

    a. 0.8 feet per second

    b. 48.9 feet per second

    c. 0.009 feet per second

    d. 70.4 feet per second

18. This chart indicates how many sales of CDs, vinyl records, and MP3 downloads occurred over the last year. Approximately what percentage of the total sales was from CDs?

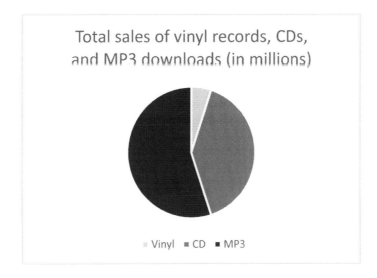

    a. 55%

    b. 25%

    c. 40%

    d. 5%

19. After a 20% sale discount, Frank purchased a new refrigerator for $850. How much did he save from the original price?

    a. $170

    b. $212.50

    c. $105.75

    d. $200

20. Which of the following is NOT a way to write 40 percent of N?
    a. $(0.4)N$

    b. $\frac{2}{5}N$

    c. $40N$

    d. $\frac{4N}{10}$

21. The graph of which function has an x-intercept of -2?
    a. $y = 2x - 3$
    b. $y = 4x + 2$
    c. $y = x^2 + 5x + 6$
    d. $y = -\frac{1}{2} \times 2^x$

22. The table below displays the number of three-year-olds at Kids First Daycare who are potty-trained and those who still wear diapers.

	Potty-trained	Wear diapers	Total
Boys	26	22	48
Girls	34	18	52
Total	60	40	

What is the probability that a three-year-old girl chosen at random from the school is potty-trained?
    a. 52 percent
    b. 34 percent
    c. 65 percent
    d. 57 percent

23. A clothing company with a target market of U.S. boys surveys 2,000 twelve-year-old boys to find their height. The average height of the boys is 61 inches. For the above scenario, 61 inches represents which of the following?
    a. Sample statistic
    b. Population parameter
    c. Confidence interval
    d. Measurement error

24. A government agency is researching the average consumer cost of gasoline throughout the United States. Which data collection method would produce the most valid results?
    a. Randomly choosing one hundred gas stations in the state of New York
    b. Randomly choosing ten gas stations from each of the fifty states
    c. Randomly choosing five hundred gas stations from across all fifty states with the number chosen proportional to the population of the state
    d. Methods A, B, and C would each produce equally valid results.

25. What is the equation of a circle whose center is (1, 5) and radius is 4?
    a. $(x - 1)^2 + (y - 25)^2 = 4$
    b. $(x - 1)^2 + (y - 25)^2 = 16$
    c. $(x + 1)^2 + (y + 5)^2 = 16$
    d. $(x - 1)^2 + (y - 5)^2 = 16$

26. What is the slope of this line?

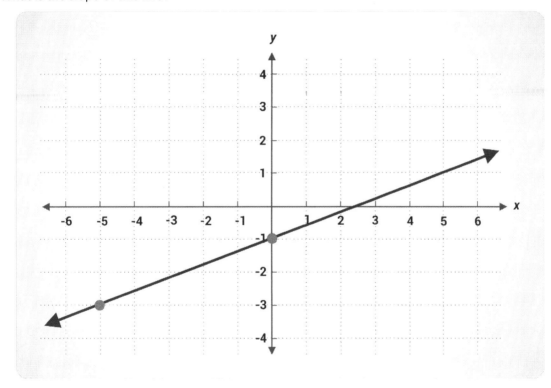

    a. 2

    b. $\frac{5}{2}$

    c. $\frac{1}{2}$

    d. $\frac{2}{5}$

27. What is the perimeter of the figure below? Note that the solid outer line is the perimeter.

8 in.

8 in.

4 in.   4 in.

a. 48.566 ft
b. 36.566 ft
c. 19.78 ft
d. 30.566 ft

28 Which of the following equations best represents the problem below?
The width of a rectangle is 2 centimeters less than the length. If the perimeter of the rectangle is 44 centimeters, then what are the dimensions of the rectangle?

a. $2l + 2(l - 2) = 44$
b. $(l + 2) + (l + 2) + l = 48$
c. $l \times (l - 2) = 44$
d. $(l + 2) + (l + 2) + l = 44$

29. How will the following algebraic expression be simplified: $(5x^2 - 3x + 4) - (2x^2 - 7)$?
a. x^5
b. $3x^2 - 3x + 11$
c. $3x^2 - 3x - 3$
d. $x - 3$

30. What is the product of the following expression?
$$(4x - 8)(5x^2 + x + 6)$$
a. $20x^3 - 36x^2 + 16x - 48$
b. $6x^3 - 41x^2 + 12x + 15$
c. $204 + 11x^2 - 37x - 12$
d. $2x^3 - 11x^2 - 32x + 20$

31. What is the solution for the following equation?
$$\frac{x^2 + x - 30}{x - 5} = 11$$
a. $x = -6$
b. There is no solution.
c. $x = 16$
d. $x = 5$

32. What are the zeros of the function: $f(x) = x^3 + 4x^2 + 4x$?

    a. -2

    b. 0, -2

    c. 2

    d. 0, 2

33. Karen gets paid a weekly salary and a commission for every sale that she makes. The table below shows the number of sales and her pay for different weeks.

Sales	2	7	4	8
Pay	$380	$580	$460	$620

Which of the following equations represents Karen's weekly pay?

    a. $y = 90x + 200$

    b. $y = 90x - 200$

    c. $y = 40x + 300$

    d. $y = 40x - 300$

34. The square and circle have the same center. The circle has a radius of r. What is the area of the shaded region?

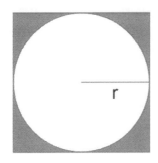

    a. $r^2 - \pi r^2$

    b. $4r^2 - 2\pi r$

    c. $(4 - \pi)r^2$

    d. $(\pi - 1)r^2$

35. For the following similar triangles, what are the values of x and y (rounded to one decimal place)?

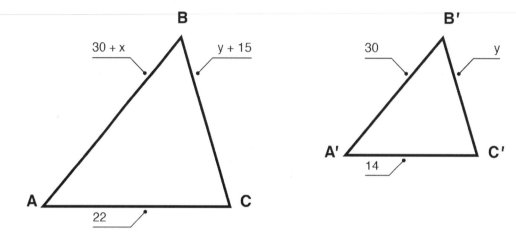

a. $x = 16.5, y = 25.1$
b. $x = 19.5, y = 24.1$
c. $x = 17.1, y = 26.3$
d. $x = 26.3, y = 17.1$

36. What are the center and radius of a circle with equation $4x^2 + 4y^2 - 16x - 24y + 51 = 0$?
a. Center (3, 2) and radius ½
b. Center (2, 3) and radius ½
c. Center (3, 2) and radius ¼
d. Center (2, 3) and radius ¼

37. Last year, the New York City area received approximately $27\frac{3}{4}$ inches of snow. The Denver area received approximately 3 times as much snow as New York City. How much snow fell in Denver?
a. 60 inches
b. $27\frac{1}{4}$ inches
c. $9\frac{1}{4}$ inches
d. $83\frac{1}{4}$ inches

38. In Jim's school, there are 3 girls for every 2 boys. There are 650 students in total. Using this information, how many students are girls?

No Calculator Questions

39. If $-3(x + 4) \geq x + 8$, what is the value of x?
   a. $x = 4$
   b. $x \geq 2$
   c. $x \geq -5$
   d. $x \leq -5$

40. Kristen purchases $100 worth of CDs and DVDs. The CDs cost $10 each and the DVDs cost $15. If she bought four DVDs, how many CDs did she buy?

41. A student gets an 85% on a test with 20 questions. How many answers did the student solve correctly?

42. Four people split a bill. The first person pays for $\frac{1}{5}$, the second person pays for $\frac{1}{4}$, and the third person pays for $\frac{1}{3}$. What fraction of the bill does the fourth person pay?

    a. $\frac{13}{60}$

    b. $\frac{47}{60}$

    c. $\frac{1}{4}$

    d. $\frac{4}{15}$

43. Five of six numbers have a sum of 25. The average of all six numbers is 6. What is the sixth number?

44. Kimberley earns $10 an hour babysitting, and after 10 p.m., she earns $12 an hour, with the amount paid being rounded to the nearest hour accordingly. On her last job, she worked from 5:30 p.m. to 11 p.m. In total, how much did Kimberley earn for that job?

a. $45
b. $57
c. $62
d. $42

45. A family purchases a vehicle in 2005 for $20,000. In 2010, they decide to sell it for a newer model. They are able to sell the car for $8,000. By what percentage did the value of the family's car drop?

a. 40%
b. 68%
c. 60%
d. 33%

46. In May of 2010, a couple purchased a house for $100,000. In September of 2016, the couple sold the house for $93,000 so they could purchase a bigger one to start a family. How many months did they own the house?

a. 76
b. 54
c. 85
d. 93

47. At the beginning of the day, Xavier has 20 apples. At lunch, he meets his sister Emma and gives her half of his apples. After lunch, he stops by his neighbor Jim's house and gives him 6 of his apples. He then uses $\frac{3}{4}$ of his remaining apples to make an apple pie for dessert at dinner. At the end of the day, how many apples does Xavier have left?

    a. 4
    b. 6
    c. 2
    d. 1

48. If $\frac{5}{2} \div \frac{1}{3} = n$, then n is between:

    a. 5 and 7
    b. 7 and 9
    c. 9 and 11
    d. 3 and 5

49. A closet is filled with red, blue, and green shirts. If $\frac{1}{3}$ of the shirts are green and $\frac{2}{5}$ are red, what fraction of the shirts are blue?

    a. $\frac{4}{15}$

    b. $\frac{1}{5}$

    c. $\frac{7}{15}$

    d. $\frac{1}{2}$

50. Shawna buys $2\frac{1}{2}$ gallons of paint. If she uses $\frac{1}{3}$ of it on the first day, how much does she have left?

    a. $1\frac{5}{6}$ gallons

    b. $1\frac{1}{2}$ gallons

    c. $1\frac{2}{3}$ gallons

    d. 2 gallons

51. What is the volume of a cylinder, in terms of π, with a radius of 6 centimeters and a height of 2 centimeters?

    a. 36π cm³
    b. 24π cm³
    c. 72π cm³
    d. 48π cm³

52. Keith's bakery had 252 customers go through its doors last week. This week, that number increased to 378. By what percentage did his customer volume increase?

    a. 26%

    b. 50%

    c. 35%

    d. 12%

53. Twenty is 40% of what number?

    a. 50

    b. 8

    c. 200

    d. 5000

54. If Danny takes 48 minutes to walk 3 miles, how many minutes should it take him to walk 5 miles maintaining the same speed?

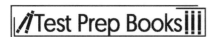

55. The perimeter of a 6-sided polygon is 56 cm. The length of three sides are 9 cm each. The length of two other sides are 8 cm each. What is the length of the missing side?

56. If the sine of $30° = x$, the cosine of what angle, in degrees, also equals x?

57. What is the value of $x^2 - 2xy + 2y^2$ when $x = 2, y = 3$?

58. If $4x - 3 = 5$, then what is the value of x?

Essay Prompt

Bob's lawnmower shop has an excess supply of lawnmowers at the end of the summer season for the first time in years. Bob's son reasons that this excess supply should be heavily discounted and sold as quickly as possible to make room for winter equipment such as snowblowers and snowmobiles to maximize profits and revenue. Bob wants to keep the unsold lawnmowers and attempt to sell them again next year when demand returns.

Write a response in which you discuss what questions should be asked and resolved in order to validate Bob's strategy as a valid one versus his son's idea. Be sure to explain how easy those answers are to obtain, what assumptions they involve, and how the answers to the questions support Bob's strategy and confirm that it has a high probability of success.

Answer Explanations #2

Reading Test

1. D: To define and describe instances of spinoff technology. This is an example of a purpose question—*why* did the author write this? The article contains facts, definitions, and other objective information without telling a story or arguing an opinion. In this case, the purpose of the article is to inform the reader. The only choice that is related to giving information is Choice *D*: to define and describe instances of spinoff technology.

2. A: A general definition followed by more specific examples. This organization question asks readers to analyze the structure of the essay. The topic of the essay is about spinoff technology; the first paragraph gives a general definition of the concept, while the following two paragraphs offer more detailed examples to help illustrate this idea.

3. C: They were looking for ways to add health benefits to food. This reading comprehension question can be answered based on the second paragraph—scientists were concerned about astronauts' nutrition and began researching useful nutritional supplements. Choice *A* in particular is not true because it reverses the order of discovery (first NASA identified algae for astronaut use, and then it was further developed for use in baby food).

4. B: Related to the brain. This vocabulary question could be answered based on the reader's prior knowledge; but even for readers who have never encountered the word "neurological" before, the passage does provide context clues. The very next sentence talks about "this algae's potential to boost brain health," which is a paraphrase of "neurological benefits." From this context, readers should be able to infer that "neurological" is related to the brain.

5. D: To give an example of valuable space equipment. This purpose question requires readers to understand the relevance of the given detail. In this case, the author mentions "costly and crucial equipment" before mentioning space suit visors, which are given as an example of something that is very valuable. Choice *A* is not correct because fashion is only related to sunglasses, not to NASA equipment. Choice *B* can be eliminated because it is simply not mentioned in the passage. While Choice *C* seems like it could be a true statement, it is also not relevant to what is being explained by the author.

6. C: It is difficult to make money from scientific research. The article gives several examples of how businesses have been able to capitalize on NASA research, so it is unlikely that the author would agree with this statement. Evidence for the other answer choices can be found in the article: Choice *A*, the author mentions that "many consumers are unaware that products they are buying are based on NASA research"; Choice *B* is a general definition of spinoff technology; and Choice *D* is mentioned in the final paragraph.

7. D: To criticize a theory by presenting counterevidence. The author mentions anti-Stratfordian arguments in the first paragraph, but then goes on to debunk these theories with more facts about Shakespeare's life in the second and third paragraphs. Choice *A* is not correct because, while the author does present arguments from both sides, the author is far from unbiased; in fact, the author clearly disagrees with anti-Stratfordians. Choice *B* is also not correct because it is more closely aligned to the beliefs of anti-Stratfordians, whom the author disagrees with. Choice *C* can be eliminated because, while

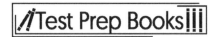

it is true that the author gives historical background, the main purpose of the article is using that information to disprove a theory.

8. B: But in fact, there is not much substance to such speculation, and most anti-Stratfordian arguments can be refuted with a little background about Shakespeare's time and upbringing. The thesis is a statement that contains the author's topic and main idea. The main purpose of this article is to use historical evidence to provide counterarguments to anti-Stratfordians. Choice *A* is simply a definition; Choice *C* is a supporting detail, not a main idea; and Choice *D* represents an idea of anti-Stratfordians, not the author's opinion.

9. C: Rhetorical question. This requires readers to be familiar with different types of rhetorical devices. A rhetorical question is a question that is asked not to obtain an answer but to encourage readers to more deeply consider an issue.

10. B: By explaining grade school curriculum in Shakespeare's time. This question asks readers to refer to the organizational structure of the article and demonstrate understanding of how the author provides details to support their argument. This particular detail can be found in the second paragraph: "even though he did not attend university, grade school education in Shakespeare's time was actually quite rigorous."

11. A: Busy. This is a vocabulary question that can be answered using context clues. Other sentences in the paragraph describe London as "the most populous city in England" filled with "crowds of people," giving an image of a busy city full of people. Choice *B* is not correct because London was in Shakespeare's home country, not a foreign one. Choice *C* is not mentioned in the passage. Choice *D* is not a good answer choice because the passage describes how London was a popular and important city, probably not an undeveloped one.

12. D: In Shakespeare's time, glove-makers were not part of the upper class. Anti-Stratfordians doubt Shakespeare's ability because he was not from the upper class; his father was a glove-maker; therefore, in at least this instance, glove-makers were not included in the upper class (this is an example of inductive reasoning, or using two specific pieces of information to draw a more general conclusion).

13. C: It is an example of a play set outside of England. This detail comes from the third paragraph, where the author responds to skeptics who claim that Shakespeare wrote too much about places he never visited, so *Romeo and Juliet* is mentioned as a famous example of a play with a foreign setting. In order to answer this question, readers need to understand the author's main purpose in the third paragraph and how the author uses details to support this purpose. Choices *A* and *D* are not mentioned in the passage, and Choice *B* is clearly not true because the passage mentions more than once that Shakespeare never left England.

14. A: It is possible to learn things from reading rather than from firsthand experience. This inference can be made from the final paragraph, where the author refutes anti-Stratfordian skepticism by pointing out that books about life in Europe could easily circulate throughout London. From this statement, readers can conclude that the author believes it is possible that Shakespeare learned about European culture from books, rather than visiting the continent on his own. Choice *B* is not true because the author believes that Shakespeare contributed to English literature without traveling extensively. Similarly, Choice *C* is not a good answer because the author explains how Shakespeare got his education without university. Choice *D* can also be eliminated because the author describes Shakespeare's genius and clearly Shakespeare is not from Italy.

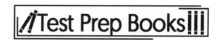

15. A: First person. This is a straightforward question that requires readers to know that a first-person narrator speaks from an "I" point of view.

16. D: He doesn't understand all of the languages being used. This can be inferred from the fact that the traveler must refer to his dictionary to understand those around him. Choice *A* isn't a good choice because the traveler seems to wonder why the driver needs to drive so fast. Choice *B* isn't mentioned in the passage and doesn't seem like a good answer choice because he seems wholly unfamiliar with his surroundings. This is why Choice *C* can also be eliminated.

17. B: From fearful to charmed. This can be found in the first sentence of the third paragraph, which states, "I soon lost sight and recollection of ghostly fears in the beauty of the scene as we drove along." Also, readers should get a sense of foreboding from the first two paragraphs, where superstitious villagers seem frightened on the traveler's behalf. However, the final paragraph changes to delighted descriptions of the landscape's natural beauty. Choices *A* and *D* can be eliminated because the traveler is anxious, not relaxed or comfortable at the beginning of the passage. Choice *C* can also be eliminated because the traveler doesn't gain any particular insights in the last paragraph, and in fact continues to lament that he cannot understand the speech of those around him.

18. D: A complete stranger. The answer to this reading comprehension question can be found in the second paragraph, when the traveler is "just starting for an unknown place to meet an unknown man"— In other words, a complete stranger.

19. C: The traveler will soon encounter danger or evil. Answering this prediction question requires readers to understand foreshadowing, or hints that the author gives about what will happen next. There are numerous hints scattered throughout this passage: the villager's sorrow and sympathy for the traveler and their superstitious actions; the spooky words that the traveler overhears; the driver's unexplained haste. All of these point to a danger that awaits the protagonist.

20. A: "I must say they weren't cheering to me, for amongst them were "Ordog"—Satan, "pokol"—hell, "stregoica"—witch, "vrolok" and "vlkoslak"—both of which mean the same thing, one being Slovak and the other Servian for something that is either were-wolf or vampire." This sentence is an example of how the author hints at evil to come for the traveler. The other answer choices aren't related to the passage's grim foreshadowing.

21. D: It emphasizes Mr. Utterson's anguish in failing to identify Hyde's whereabouts. Context clues indicate that Choice *D* is correct because the passage provides great detail of Mr. Utterson's feelings about locating Hyde. Choice *A* does not fit because there is no mention of Mr. Lanyon's mental state. Choice *B* is incorrect; although the text does make mention of bells, Choice *B* is not the *best* answer overall. Choice *C* is incorrect because the passage clearly states that Mr. Utterson was determined, not unsure.

22. A: In the city. The word *city* appears in the passage several times, thus establishing the location for the reader.

23. B: It scares children. The passage states that the Juggernaut causes the children to scream. Choices *A* and *D* don't apply because the text doesn't mention either of these instances specifically. Choice *C* is incorrect because there is nothing in the text that mentions space travel.

24. B: To constantly visit. The mention of *morning*, *noon*, and *night* make it clear that the word *haunt* refers to frequent appearances at various times. Choice *A* doesn't work because the text

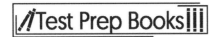

makes no mention of levitating. Choices *C* and *D* are not correct because the text makes mention of Mr. Utterson's anguish and disheartenment because of his failure to find Hyde but does not make mention of Mr. Utterson's feelings negatively affecting anyone else.

25. D: This is an example of alliteration. Choice *D* is the correct answer because of the repetition of the *L*-words. Hyperbole is an exaggeration, so Choice *A* doesn't work. No comparison is being made, so no simile or metaphor is being used, thus eliminating Choices *B* and *C*.

26. D: The speaker intends to continue to look for Hyde. Choices *A* and *B* are not possible answers because the text doesn't refer to any name changes or an identity crisis, despite Mr. Utterson's extreme obsession with finding Hyde. The text also makes no mention of a mistaken identity when referring to Hyde, so Choice *C* is also incorrect.

27. C: The author contrasts two different viewpoints, then builds a case showing preference for one over the other. Choice *A* is incorrect because the introduction does not contain an impartial definition, but rather, an opinion. Choice *B* is incorrect. There is no puzzling phenomenon given, as the author doesn't mention any peculiar cause or effect that is in question regarding poetry. Choice *D* does contain another's viewpoint at the beginning of the passage; however, to say that the author has no stake in this argument is incorrect; the author uses personal experiences to build their case.

28. B: Choice *B* accurately describes the author's argument in the text: that poetry is not irrelevant. While the author does praise, and even value, Buddy Wakefield as a poet, the author never heralds him as a genius. Eliminate Choice *A*, as it is an exaggeration. Not only is Choice *C* an exaggerated statement, but the author never mentions spoken word poetry in the text. Choice *D* is wrong because this statement contradicts the writer's argument.

29. D: *Exiguously* means not occurring often, or occurring rarely, so Choice *D* would LEAST change the meaning of the sentence. Choice *A*, *indolently*, means unhurriedly, or slow, and does not fit the context of the sentence. Choice *B*, *inaudibly*, means quietly or silently. Choice *C*, *interminably*, means endlessly, or all the time, and is the opposite of the word *exiguously*.

30. D: A student's insistence that psychoanalysis is a subset of modern psychology is the most analogous option. The author of the passage tries to insist that performance poetry is a subset of modern poetry, and therefore, tries to prove that modern poetry is not "dying," but thriving on social media for the masses. Choice *A* is incorrect, as the author is not refusing any kind of validation. Choice *B* is incorrect; the author's insistence is that poetry will *not* lose popularity. Choice *C* mimics the topic but compares two different genres, while the author does no comparison in this passage.

31. B: The author's purpose is to disprove Gioia's article claiming that poetry is a dying art form that only survives in academic settings. In order to prove his argument, the author educates the reader about new developments in poetry (Choice *A*) and describes the brilliance of a specific modern poet (Choice *C*), but these serve as examples of a growing poetry trend that counters Gioia's argument. Choice *D* is incorrect because it contradicts the author's argument.

32. D: This question is difficult because the choices offer real reasons as to why the author includes the quote. However, the question specifically asks for the *main reason* for including the quote. The quote from a recently written poem shows that people are indeed writing, publishing, and performing poetry (Choice *B*). The quote also shows that people are still listening to poetry (Choice *C*). These things are true, and by their nature, serve to disprove Gioia's views (Choice *A*), which is the author's goal.

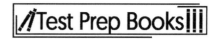

However, Choice *D* is the most direct reason for including the quote, because the article analyzes the quote for its "complex themes" that "draws listeners and appreciation" right after it's given.

33. B: Mr. Button's wife is about to have a baby. The passage begins by giving the reader information about traditional birthing situations. Then, we are told that Mr. and Mrs. Button decide to go against tradition to have their baby in a hospital. The next few passages are dedicated to letting the reader know how Mr. Button dresses and goes to the hospital to welcome his new baby. There is a doctor in this excerpt, as Choice *C* indicates, and Mr. Button does put on clothes, as Choice *D* indicates. However, Mr. Button is not going to the doctor's office nor is he about to go shopping for new clothes.

34. A: The tone of the above passage is nervous and excited. We are told in the fourth paragraph that Mr. Button "arose nervously." We also see him running without caution to the doctor to find out about his wife and baby—this indicates his excitement. We also see him stuttering in a nervous yet excited fashion as he asks the doctor if it's a boy or girl. Though the doctor may seem a bit abrupt at the end, indicating a bit of anger or shame, neither of these choices is the overwhelming tone of the entire passage. Despite the circumstances, joy and gratitude are not the main tone in the passage.

35. C: Dedicated. Mr. Button is dedicated to the task before him. Choice *A*, *numbed*, Choice *B*, *chained*, and Choice *D*, *moved*, all could grammatically fit in the sentence. However, they are not synonyms with *consecrated* like Choice *C* is.

36. D: Giving readers a visual picture of what the doctor is doing. The author describes a visual image—the doctor rubbing his hands together—first and foremost. The author may be trying to make a comment about the profession; however, the author does not "explain the detail of the doctor's profession" as Choice *B* suggests.

37. D: It introduces the setting of the story and its characters. We know we are being introduced to the setting because we are given the year in the very first paragraph along with the season: "one day in the summer of 1860." This is a classic structure of an introduction of the setting. We are also getting a long explanation of Mr. Button, what his work is, who is related to him, and what his life is like in the third paragraph.

38. B: "Talk sense!" is an example of an imperative sentence. An imperative sentence gives a command. The doctor is commanding Mr. Button to talk sense. Choice *A* is an example of an exclamatory sentence, which expresses excitement. Choice *C* is an example of an interrogative sentence—this type of sentence asks questions. Choice *D* is an example of a declarative sentence. This means that the character is simply making a statement.

39. C: The point of view is told in third-person omniscient. We know this because the story starts out with us knowing something that the character does not know: that her husband has died. Mrs. Mallard eventually comes to know this, but we as readers know this information before it is broken to her. In third person limited, Choice *D*, we would only see and know what Mrs. Mallard herself knew, and we would find out the news of her husband's death when she found out the news, not before.

40. A: The way Mrs. Mallard reacted to her husband's death. The irony in this story is called situational irony, which means the situation that takes place is different than what the audience anticipated. At the beginning of the story, we see Mrs. Mallard react with a burst of grief to her husband's death. However, once she's alone, she begins to contemplate her future and says the word "free" over and over. This is quite a different reaction from Mrs. Mallard than what readers expected from the start of the story.

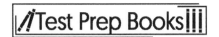

41. B: The word *elusive* most closely means *indefinable*. *Horrible*, Choice *A*, doesn't quite fit with the tone of the word *subtle* that comes before it. Choice *C*, *quiet*, is more closely related to the word *subtle*. Choice *D*, *joyful*, also doesn't quite fit the context here. *Indefinable* is the best option.

42. D: Mrs. Mallard, a newly widowed woman, finds unexpected relief in her husband's death. A summary is a brief explanation of the main point of a story. The story mostly focuses on Mrs. Mallard and her reaction to her husband's death, especially in the room when she's alone and contemplating the present and future. All of the other answer choices except Choice *C* are briefly mentioned in the story; however, they are not the main focus of the story.

43. D: The interesting thing about this story is that feelings that are confused, joyful, and depressive all play a unique and almost equal part of this story. There is no one right answer here, because the author seems to display all of these emotions through the character of Mrs. Mallard. She displays feelings of depressiveness by her grief at the beginning; then, when she receives feelings of joy, she feels moments of confusion. We as readers cannot help but go through these feelings with the character. Thus, the author creates a tone of depression, joy, and confusion, all in one story.

44. C: The word *tumultuously* most nearly means *violently*. Even if you don't know the word *tumultuously*, look at the surrounding context to figure it out. The next few sentences we see Mrs. Mallard striving to "beat back" the "thing that was approaching to possess her." We see a fearful and almost violent reaction to the emotion that she's having. Thus, her chest would rise and fall tumultuously, or violently.

45. A: The purpose is to inform the reader about what assault is and how it is committed. Choice *B* is incorrect because the passage does not state that assault is a lesser form of lethal force, only that an assault can use lethal force, or alternatively, lethal force can be utilized to counter a dangerous assault. Choice *C* is incorrect because the passage is informative and does not have a set agenda. Finally, Choice *D* is incorrect because although the author uses an example in order to explain assault, it is not indicated that this is the author's personal account.

46. C: If the man being attacked in an alley by another man with a knife used self-defense by lethal force, it would not be considered illegal. The presence of a deadly weapon indicates mal-intent and because the individual is isolated in an alley, lethal force in self-defense may be the only way to preserve his life. Choices *A* and *B* can be ruled out because in these situations, no one is in danger of immediate death or bodily harm by someone else. Choice *D* is an assault and does exhibit intent to harm, but this situation isn't severe enough to merit lethal force; there is no intent to kill.

47. B: As discussed in the second passage, there are several forms of assault, like assault with a deadly weapon, verbal assault, or threatening posture or language. Choice *A* is incorrect because the author does mention what the charges are on assaults; therefore, we cannot assume that they are more or less than unnecessary use of force charges. Choice *C* is incorrect because anyone is capable of assault; the author does not state that one group of people cannot commit assault. Choice *D* is incorrect because assault is never justified. Self-defense resulting in lethal force can be justified.

48. D: The use of lethal force is not evaluated on the intent of the user, but rather on the severity of the primary attack that warranted self-defense. This statement most undermines the last part of the passage because it directly contradicts how the law evaluates the use of lethal force. Choices *A* and *B* are stated in the paragraph, so they do not undermine the explanation from the author. Choice *C* does not necessarily undermine the passage, but it does not support the passage either. It is more of an opinion that does not offer strength or weakness to the explanation.

49. C: An assault with deadly intent can lead to an individual using lethal force to preserve their well-being. Choice *C* is correct because it clearly establishes what both assault and lethal force are and gives the specific way in which the two concepts meet. Choice *A* is incorrect because lethal force doesn't necessarily result in assault. This is also why Choice *B* is incorrect. Not all assaults would necessarily be life-threatening to the point where lethal force is needed for self-defense. Choice *D* is compelling but ultimately too vague; the statement touches on aspects of the two ideas but fails to present the concrete way in which the two are connected to each other.

50. A: Both passages open by defining a legal concept and then continue to describe situations in order to further explain the concept. Choice *D* is incorrect because while the passages utilize examples to help explain the concepts discussed, the author doesn't indicate that they are specific court cases. It's also clear that the passages don't open with examples, but instead, they begin by defining the terms addressed in each passage. This eliminates Choice *B,* and ultimately reveals Choice *A* to be the correct answer. Choice *A* accurately outlines the way both passages are structured. Because the passages follow a nearly identical structure, the Choice *C* can easily be ruled out.

51. C: Intent is very important for determining both lethal force and assault; intent is examined in both parties and helps determine the severity of the issue. Choices *A* and *B* are incorrect because it is clear in both passages that intent is a prevailing theme in both lethal force and assault. Choice *D* is compelling, but if a person uses lethal force to defend himself or herself, the intent of the defender is also examined in order to help determine if there was excessive force used. Choice *C* is correct because it states that intent is important for determining both lethal force and assault, and that intent is used to gauge the severity of the issues. Remember, just as lethal force can escalate to excessive use of force, there are different kinds of assault. Intent dictates several different forms of assault.

52. B: The example is used to demonstrate a single example of two different types of assault, then adding in a third type of assault to the example's conclusion. The example mainly provides an instance of "threatening body language" and "provocative language" with the homeowner gesturing threats to his neighbor. It ends the example by adding a third type of assault: physical strikes. This example is used to show the variant nature of assaults. Choice *A* is incorrect because it doesn't mention the "physical strike" assault at the end and is not specific enough. Choice *C* is incorrect because the example does not say anything about the definition of lethal force or how it might be altered. Choice *D* is incorrect, as the example mentions nothing about cause and effect.

Writing and Language Test

1. B: Choice *B* is correct. Here, a colon is used to introduce an explanation. Colons either introduce explanations or lists. Additionally, the quote ends with the punctuation inside the quotes, unlike Choice *C.*

2. A: The verb tense in this passage is predominantly in the present tense, so Choice *A* is the correct answer. Choice *B* is incorrect because the subject and verb do not agree. It should be "Education provides," not "Education provide." Choice *C* is incorrect because the passage is in present tense, and "Education will provide" is future tense. Choice *D* doesn't make sense when placed in the sentence.

3. D: The possessive form of the word "it" is "its." The contraction "it's" denotes "it is." Thus, Choice *A* is wrong. The word "raises" in Choice *B* makes the sentence grammatically incorrect. Choice *C* adds an apostrophe at the end of "its." While adding an apostrophe to most words would indicate possession, adding *'s* to the word "it" indicates a contraction.

4. C: The word *civilised* should be spelled *civilized*. The words *distinguishes* and *creatures* are both spelled correctly.

5. B: Choice *B* is correct because it provides clarity by describing what "myopic" means right after the word itself. Choice *A* is incorrect because the explanation of "myopic" comes before the word; thus, the meaning is skewed. It's possible that Choice *C* makes sense within context. However, it's not the *best* way to say this. Choice *D* is confusingly worded. Using "myopic focus" is not detrimental to society; however, the way *D* is worded makes it seem that way.

6. C: Again, we see where the second paragraph can be divided into two parts due to separate topics. The first section's main focus is education addressing the mind, body, and soul. The first section, then, could end with the concluding sentence, "The human heart and psyche . . ." The next sentence to start a new paragraph would be "Education is a basic human right." The rest of this paragraph talks about what education is and some of its characteristics.

7. A: Choice *A* is correct because the phrase "others' ideas" is both plural and indicates possession. Choice *B* is incorrect because "other's" indicates only one "other" that's in possession of "ideas," which is incorrect. Choice *C* is incorrect because no possession is indicated. Choice *D* is incorrect because the word *other* does not end in *s*. *Others's* is not a correct form of the word in any situation.

8. D: This sentence must have a comma before "although" because the word "although" is connecting two independent clauses. Thus, Choices *B* and *C* are incorrect. Choice *A* is incorrect because the second sentence in the underlined section is a fragment.

9. C: Choice *C* is the correct choice because the word "their" indicates possession, and the text is talking about "their students," or the students of someone. Choice *A*, "there," means at a certain place and is incorrect. Choice *B*, "they're," is a contraction and means "they are." Choice *D* is not a word.

10. B: Choice *B* uses all punctuation correctly in this sentence. In American English, single quotes should only be used if they are quotes within a quote, making Choices *A* and *C* incorrect. Additionally, punctuation here should go inside the quotes, making Choice *D* incorrect.

11. B: Choice *B* is correct because the conjunction *and* is used to connect phrases that are to be used jointly, such as teachers working hard to help students "identify salient information" and to "think critically." The conjunctions *so*, *but*, and *nor* are incorrect in the context of this sentence.

12. A: Choice *A* has consistent parallel structure with the verbs "read," "identify," and "determine." Choices *B* and *C* have faulty parallel structure with the words "determining" and "identifying." Choice *D* has incorrect subject/verb agreement. The sentence should read, "Students have to read . . . identify . . . and determine."

13. D: The correct choice for this sentence is that "they are . . . shaped by the influences." The prepositions "for," "to," and "with" do not make sense in this context. People are *shaped by*, not *shaped for, shaped to,* or *shaped with*.

14. A: To see which answer is correct, it might help to place the subject, "Teachers," near the verb. Choice *A* is correct: "Teachers . . . must strive" makes grammatical sense here. Choice *B* is incorrect because "Teachers . . . to strive" does not make grammatical sense. Choice *C* is incorrect because "Teachers must not only respect . . . but striving" eschews parallel structure. Choice *D* is incorrect because it is in past tense, and this passage is in present tense.

15. C: Choice *C* is correct because it uses an em-dash. Em-dashes are versatile. They can separate phrases that would otherwise be in parenthesis, or they can stand in for a colon. In this case, a colon would be another decent choice for this punctuation mark because the second sentence expands upon the first sentence. Choice *A* is incorrect because the statement is not a question. Choice *B* is incorrect because adding a comma here would create a comma splice. Choice *D* is incorrect because this creates a run-on sentence since the sentence contains two independent clauses.

16. B: Choice *B* is correct because it removes the apostrophe from *icon's*, since the noun *icon* is not possessing anything. This conveys the author's intent of setting *Frankenstein* apart from other icons of the romantic and science fiction genres. Choices *A* and *C* are therefore incorrect. Choice *D* is a good revision but alters the meaning of the sentence—*Frankenstein* is one of the icons, not the sole icon.

17 C: Choice *C* correctly adds a comma after *style*, successfully joining the dependent and the independent clauses as a single sentence. Choice *A* is incorrect because the dependent and independent clauses remain unsuccessfully combined without the comma. Choices *B* and *D* do nothing to fix this.

18. A: Choice *A* is correct, as the sentence doesn't require changes. Choice *B* incorrectly changes the noun *enlightenment* into the verb *enlighten*. Choices *C* and *D* alter the original meaning of the sentence.

19. B: Choice *B* is correct, fixing the incorrect split of *highlight*. This is a polyseme, a word combined from two unrelated words to make a new word. On their own, *high* and *light* make no sense for the sentence, making Choice *A* incorrect. Choice *C* incorrectly decapitalizes *Ancient*—since it modifies *Greece* and works with the noun to describe a civilization, *Ancient Greece* functions as a proper noun, which should be capitalized. Choice *D* uses *highlighting*, a gerund, but the present tense of *highlight* is what works with the rest of the sentence; to make this change, a comma would be needed after *Rome*.

20. A: Choice *A* is correct, as *not only* and *but also* are correlative pairs. In this sentence, *but* successfully transitions the first part into the second half, making punctuation unnecessary. Additionally, the use of *to* indicates that an idea or challenge is being presented to the reader. Choices *B*, *C*, and *D* are not as active, meaning these revisions weaken the sentence.

21. D: Choice *D* is correct, adding finer details to help the reader understand exactly what Prometheus did and his impact: fire came with knowledge and power. Choice *A* lacks a comma after *fire*. Choice *B* inserts unnecessary commas since *people* is not part of the list *knowledge and power*. Choice *C* is a strong revision but could be confusing, hinting that the fire was knowledge and power itself, as opposed to being symbolized by the fire.

22. C: Choice *C* reverses the order of the section, making the sentence more direct. Choice *A* lacks a comma after *gods*, and although Choice *B* adds this, the structure is too different from the first half of the sentence to flow correctly. Choice *D* is overly complicated and repetitious in its structure even though it doesn't need any punctuation.

23. B: Choice *B* fixes the two problems of the sentence, changing *faltered* to present tense in agreement with the rest of the passage, and correctly linking the two dependent clauses. Choice *A* is therefore incorrect. Choice *C* does not correct the past tense of *faltered*. Choice *D* correctly adds the conjunction *by*, but it lacks a comma after the conjunction *yet*.

24. C: Choice *C* successfully applies a comma after *power*, distinguishing the causes of Frankenstein's suffering and maintaining parallel structure. Choice *A* is thus incorrect. Choice *B* lacks the necessary punctuation and unnecessarily changes *lack* to a gerund. Choice *D* is unnecessarily wordy, making the sentence more cumbersome.

25. D: Choice *D* correctly inserts an apostrophe into the contraction *doesn't*. Choice *A* is incorrect because of this omission. Choices *B* and *C* are better than the original but do not fit well with the informal tone of the passage.

26. B: Choice *B* is correct, successfully combining the two independent clauses of this compound sentence by adding a comma before *and* to create the effective pause and transition between clauses. Choice *A* does not join the independent clauses correctly. Choices *C* and *D* offer alternate ways of joining these clauses, but since *and* is already part of the sentence, adding the comma is the most logical choice. This also keeps the informal tone set by the rest of the passage.

27. C: Choice *C* correctly fixes the homophone issue of *their* and *they're*. *Their* implies ownership, which is not needed here. The author intends *they're*, a contraction of *they are*. Thus, Choices *A* and *B* are incorrect because they use the homophone *there*. Choice *D* eliminates the homophone issue altogether, but the sentence becomes more clunky because of that.

28. B: Choice *B* correctly joins the two independent clauses with a comma before *but*. Choice *A* is incorrect because, without the comma, it is a run-on sentence. Choice *C* also lacks punctuation and uses *however*, which should be reserved for starting a new sentence or perhaps after a semicolon. Choice *D* is incorrect because the semicolon throws off the sentence structure and is incorrectly used; the correct revision would have also removed *but*.

29. A: Choice *A* is correct because the sentence does not require modification. Choice *B* is incorrect because it uses the faulty subject-verb agreement, "Physical proof are." Choice *C* is incorrect because a comma would need to follow *exists*. Choice *D* is incorrect because the comma after *science* is unnecessary.

30. D: Choice *D* correctly changes *this* to *the* and retains *validity*, making it the right choice. Choices *A* and *B* keep *this*, which is not as specific as *the*. Choice *C* incorrectly pluralizes *validity*.

31. A: Choice *A* is correct because the sentence is fine without revisions. Choice *B* is incorrect, since removing *there is* is unnecessary and confusing. Choice *C* is incorrect since it inserts an unnecessary comma. Choice *D* introduces a homophone issue: *weather* refers to climatic states and atmospheric events, while *whether* expresses doubt, which is the author's intent.

32. C: Choice *C* correctly changes *continuing* to the present tense. Choice *A* is incorrect because of this out-of-place gerund use. Choice *B* not only does not fix this issue but also incorrectly changes *describe* into a gerund. While Choice *D* correctly uses *continue*, *describe* is incorrectly put in the past tense.

33. D: Choice *D* is correct, since it eliminates the unnecessary *as well* and adds a comma to separate the given example, making the sentence more direct. Choice *A* seems repetitive with *as well*, since it has *including*, and at the least needs punctuation. Choice *B* is poorly constructed, taking out the clearer *including*. Choice *C* also makes little sense.

34. B: Move the sentence so that it comes before the preceding sentence. For this question, place the underlined sentence in each prospective choice's position. To keep it as-is is incorrect because the father "going crazy" doesn't logically follow the fact that he was a "city slicker." Choice *C* is incorrect because the sentence in question is not a concluding sentence and does not transition smoothly into the second paragraph. Choice *D* is incorrect because the sentence doesn't necessarily need to be omitted since it logically follows the very first sentence in the passage.

35. D: Choice *D* is correct because "As it turns out" indicates a contrast from the previous sentiment, that the RV was a great purchase. Choice *A* is incorrect because the sentence needs an effective transition from the paragraph before. Choice *B* is incorrect because the text indicates it *is* surprising that the RV was a great purchase because the author was skeptical beforehand. Choice *C* is incorrect because the transition "Furthermore" does not indicate a contrast.

36. B: This sentence calls for parallel structure. Choice *B* is correct because the verbs "wake," "eat," and "break" are consistent in tense and parts of speech. Choice *A* is incorrect because the words "wake" and "eat" are present tense while the word "broke" is in past tense. Choice *C* is incorrect because this turns the sentence into a question, which doesn't make sense within the context. Choice *D* is incorrect because it breaks tense with the rest of the passage. "Waking," "eating," and "breaking" are all present participles, and the context around the sentence is in past tense.

37. C: Choice *C* is correct because it is clear and fits within the context of the passage. Choice *A* is incorrect because "We rejoiced as 'hackers'" does not give a reason why hacking was rejoiced. Choice *B* is incorrect because it does not mention a solution being found and is therefore not specific enough. Choice *D* is incorrect because the meaning is eschewed by the helping verb "had to rejoice," and the sentence suggests that rejoicing was necessary to "hack" a solution.

38. A: The original sentence is correct because the verb tense as well as the order of the sentence makes sense in the given context. Choice *B* is incorrect because the order of the words makes the sentence more confusing than it otherwise would be. Choice *C* is incorrect because "We are even making" is in present tense. Choice *D* is incorrect because "We will make" is future tense. The surrounding text of the sentence is in past tense.

39. B: Choice B is correct because there is no punctuation needed if a dependent clause ("while traveling across America") is located behind the independent clause ("it allowed us to share adventures"). Choice A is incorrect because there are two dependent clauses connected and no independent clause, and a complete sentence requires at least one independent clause. Choice C is incorrect because of the same reason as Choice A. Semicolons connect closely related independent clauses on either side of the semicolon. Choice D is incorrect because the dash simply interrupts the complete sentence.

40. C: The rule for "me" and "I" is that one should use "I" when it is the subject pronoun of a sentence, and "me" when it is the object pronoun of the sentence. Break the sentence up to see if "I" or "me" should be used. To say "Those are memories that I have now shared" makes more sense than to say "Those are memories that me have now shared." Choice *D* is incorrect because "my siblings" should come before "I."

41. B: Choice *B* is correct because it adds an apostrophe to *ones*, which indicates *one's* possession of *survival*. Choice *A* doesn't do this, so it is incorrect. This is the same for Choice *C*, but that option also takes out the crucial comma after *Ju-jitsu*. Choice *D* is incorrect because it changes *play* to *plays*. This disagrees with the plural *martial arts*, which is given by having an example of its many forms, *Ju-jitsu*. Therefore, *play* is required.

42. C: Choice *C* is the best answer because it most clearly defines the point that the author is trying to make. The original sentence would need a comma after *however* in order to continue the sentence fluidly—but this option isn't available. Choice *B* is close, but this option changes the meaning of the sentence. Therefore, the best alternative is to begin the sentence with *however* and have a comma follow right after it in order to introduce a new idea. The original context is still maintained, but the flow of the language is more streamlined. Thus, Choice *A* is incorrect. Choice *D* would need a comma before and after *however*, so it is also incorrect.

43. A: Choice *A* is the best answer for several reasons. To begin, the section is grammatically correct in using a semicolon to connect the two independent clauses. This allows the two ideas to be connected without separating them. In this context, the semicolon makes more sense for the overall sentence structure and passage as a whole. Choice *B* is incorrect because it forms a run-on. Choice *C* applies a comma in incorrect positions. Choice *D* separates the sentence in a place that does not make sense for the context.

44. D: Choice *D* is the correct answer because it fixes two key issues. First, *their* is incorrectly used. *Their* is a possessive indefinite pronoun and also an antecedent—neither of these fit the context of the sentence, so Choices *A* and *B* are incorrect. What should be used instead is *they're*, which is the contraction of *they are*, emphasizing action or the result of action in this case. Choice *D* also corrects another contraction-related issue with *its*. Again, *its* indicates possession, while *it's* is the contraction of *it is*. The latter is what's needed for the sentence to make sense and be grammatically correct. Thus, Choice *C* is also incorrect.

45. A: Choice *A* is correct because the section contains no errors and clearly communicates the writer's point. Choice *B* is incorrect because it lacks a comma after *philosophy*, needed to link the first clause with the second. Choice *C* also has this issue but additionally alters *its* to *it's*; since *it is* does not make sense in this sentence, this is incorrect. Choice *D* is incorrect because *its* is already plural possessive and does not need an apostrophe on the end.

Math Test

1. B: 300 miles in 4 hours is $\frac{300}{4} = 75$ miles per hour. In 1.5 hours, the car will go 1.5×75 miles, or 112.5 miles.

2.

One apple/orange pair costs $3 total. Therefore, Jan bought $\frac{90}{3} = 30$ total pairs, and hence, she bought 30 oranges.

3. C: Plug in the values for x and y to discover that the solution works, which is:

$$(-3)^2 + (-4)^2 = 25$$

Choices *A* and *B* are not the correct answers since the solution works. Choice *D* is not the correct answer because there is enough information to tell where the given point lies on the circle.

4.

First, the train's journey in the real world is $3 \times 50 = 150$ miles. On the map, 1 inch corresponds to 10 miles, so there is $\frac{150}{10} = 15$ inches on the map.

5. B: The total trip time is $1 + 3.5 + 0.5 = 5$ hours. The total time driving is $1 + 0.5 = 1.5$ hours. So, the fraction of time spent driving is 1.5 / 5 or 3 / 10. To get the percentage, convert this to a fraction out of 100. The numerator and denominator are multiplied by 10, with a result of 30 / 100. The percentage is the numerator in a fraction out of 100, so the answer is 30%.

6. B: The formula for the volume of a cylinder is $\pi r^2 h$, where r is the radius and h is the height. The diameter is twice the radius, so these barrels have a radius of 1 foot. That means each barrel has a volume of:

$$\pi \times 1^2 \times 3 = 3\pi \text{ cubic feet}$$

Since there are three of them, the total is $3 \times 3\pi = 9\pi$ cubic feet.

7. A: The tip is not taxed, so he pays 5% tax only on the $10.

The tax is 5% of $10, or $0.05 \times 10 = \$0.50$. Add up $10 + $2 + $0.50 to get $12.50.

8. A: The first step is to divide up $150 into four equal parts. $150 \div 4$ is 37.5, so she needs to save an average of $37.50 per day.

9.

The number of days can be found by taking the total amount Bernard needs to make and dividing it by the amount he earns per day:

$$\frac{300}{80} = \frac{30}{8} = \frac{15}{4} = 3.75$$

But Bernard is only working full days, so he will need to work 4 days, since 3 days are not sufficient.

10. A: The value went up by:

$$\$165,000 - \$150,000 = \$15,000$$

Out of \$150,000, this is $\frac{15,000}{150,000} = \frac{1}{10}$. Convert this to having a denominator of 100, which yields a result of $\frac{10}{100}$ or 10%.

11. D: The total faculty is $15 + 20 = 35$. Therefore, the faculty to student ratio is 35:200. Then, to simplify this ratio, both the numerator and the denominator are divided by 5, since 5 is a common factor of both, which yields 7:40.

12.

The journey will be $5 \times 3 = 15$ miles. A car traveling at 60 miles per hour is traveling at 1 mile per minute. The resulting equation would be:

$$\frac{15 \text{ mi}}{1 \frac{\text{mi}}{\text{min}}} = 15 \text{ min}$$

Therefore, it will take 15 minutes to make the journey.

13. A: Taylor's total income is $20,000 + $10,000 = $30,000. Fifteen percent of this is $\frac{15}{100} = \frac{3}{20}$. So:

$$\frac{3}{20} \times \$30,000 = \frac{\$90,000}{20}$$

$$\frac{\$9000}{2} = \$4500$$

14. D: Exponential functions can be written in the form: $y = a \times b^x$. The equation for an exponential function can be written given the y-intercept (a) and the growth rate (b). The y-intercept is the output (y) when the input (x) equals zero. It can be thought of as an "original value," or starting point. The value of b is the rate at which the original value increases ($b > 1$) or decreases ($b < 1$). In this scenario, the y-intercept, a, would be $1200, and the growth rate, b, would be 1.01 (100% of the original value combined with 1% interest, or $100\% + 1\% = 101\% = 1.01$).

15. C: The domain consists of all possible inputs, or x-values. The scenario states that the function approximates the population between the years 1900 and 2000. It also states that $x = 0$ represents the year 1900. Therefore, the year 2000 would be represented by $x = 100$. Only inputs between 0 and 100 are relevant in this case.

16. D: This problem can be solved by setting up a proportion involving the given information and the unknown value. The proportion is:

$$\frac{21 \text{ pages}}{4 \text{ nights}} = \frac{140 \text{ pages}}{x \text{ nights}}$$

Solving the proportion by cross-multiplying, the equation becomes $21x = 4 \times 140$, where $x = 26.67$. Since it is not an exact number of nights, the answer is rounded up to 27 nights. Twenty-six nights would not give Sarah enough time.

17. D: This problem can be solved by using unit conversion. The initial units are miles per minute. The final units need to be feet per second. Converting miles to feet uses the equivalence statement 1 mile equals 5,280 feet. Converting minutes to seconds uses the equivalence statement 1 minute equals 60 seconds. Setting up the ratios to convert the units is shown in the following equation:

$$\frac{72 \text{ mi}}{90 \text{ min}} \times \frac{1 \text{ min}}{60 \text{ s}} \times \frac{5280 \text{ ft}}{1 \text{ mi}} = 70.4 \frac{\text{ft}}{\text{s}}$$

The initial units cancel out, and the new units are left.

18. C: The sum total percentage of a pie chart must equal 100%. Since the CD sales take up less than half of the chart and more than a quarter (25%), it can be determined to be 40% overall. This can also be measured with a protractor. The angle of a circle is 360°. Since 25% of 360° would be 90° and 50% would be 180°, the angle percentage of CD sales falls in between; therefore, it would be Choice C.

19. B: Since $850 is the price *after* a 20% discount, $850 represents 80% of the original price. To determine the original price, set up a proportion with the ratio of the sale price (850) to original price (unknown) equal to the ratio of sale percentage (where x represents the unknown original price):

$$\frac{850}{x} = \frac{80}{100}$$

To solve a proportion, cross multiply the numerators and denominators and set the products equal to each other: $(850)(100) = (80)(x)$. Multiplying each side results in the equation $85,000 = 80x$.

To solve for x, divide both sides by 80: $\frac{85,000}{80} = \frac{80x}{80}$, resulting in $x = 1062.5$. Remember that x represents the original price. Subtracting the sale price from the original price ($1062.50 - $850) indicates that Frank saved $212.50.

20. C: $40N$ would be 4000% of N. It's possible to check that each of the others is actually 40% of N.

21. C: An x-intercept is the point where the graph crosses the x-axis. At this point, the value of y is 0. To determine if an equation has an x-intercept of -2, substitute -2 for x, and calculate the value of y. If the value of -2 for x corresponds with a y-value of 0, then the equation has an x-intercept of -2. The only answer choice that produces this result is Choice C:

$$0 = (-2)^2 + 5(-2) + 6$$

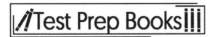

22. C: The conditional frequency of a girl being potty-trained is calculated by dividing the number of potty-trained girls by the total number of girls:

$$34 \div 52 = 0.65$$

To determine the conditional probability, multiply the conditional frequency by 100:

$$0.65 \times 100 = 65\%$$

23. A: A sample statistic indicates information about the data that was collected (in this case, the heights of those surveyed). A population parameter describes an aspect of the entire population (in this case, all twelve-year-old boys in the United States). A confidence interval would consist of a range of heights likely to include the actual population parameter. Measurement error relates to the validity of the data that was collected.

24. C: To ensure valid results, samples should be taken across the entire scope of the study. Since all states are not equally populated, representing each state proportionately would result in a more accurate statistic.

25. D: Subtract the center from the x and y values of the equation and square the radius on the right side of the equation. Choice A is not the correct answer because you need to square the radius of the equation. Choice B is not the correct answer because you do not square the centers of the equation. Choice C is not the correct answer because you need to subtract (not add) the centers of the equation.

26. D: The slope is given by the change in y divided by the change in x. Specifically, it's:

$$slope = \frac{y_2 - y_1}{x_2 - x_1}$$

The first point is (-5, -3), and the second point is (0, -1). Work from left to right when identifying coordinates. Thus the point on the left is point 1 (-5, -3) and the point on the right is point 2 (0, -1).

Now we need to just plug those numbers into the equation:

$$slope = \frac{-1 - (-3)}{0 - (-5)}$$

It can be simplified to:

$$slope = \frac{-1 + 3}{0 + 5}$$

$$slope = \frac{2}{5}$$

27. B: The figure is composed of three sides of a square and a semicircle. The sides of the square are simply added: $8 + 8 + 8 = 24$ feet. The circumference of a circle is found by the equation $C = 2\pi r$. The radius is 4, so the circumference of the circle is 25.132 ft. Only half of the circle makes up the outer border of the figure (part of the perimeter) so half of 25.132 feet is 12.566 ft. Therefore, the total perimeter is: 24 ft $+ 12.566$ ft $= 36.566$ ft. The other answer choices use the incorrect formula or fail to include all of the necessary sides.

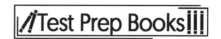

28. A: The first step is to determine the unknown, which is in terms of the length, l.

The second step is to translate the problem into the equation using the perimeter of a rectangle:

$$P = 2l + 2w$$

The width is the length minus 2 centimeters. The resulting equation is:

$$2l + 2(l - 2) = 44$$

The equation can be solved as follows:

$2l + 2l - 4 = 44$	Apply the distributive property on the left side of the equation
$4l - 4 = 44$	Combine like terms on the left side of the equation
$4l = 48$	Add 4 to both sides of the equation
$l = 12$	Divide both sides of the equation by 4

The length of the rectangle is 12 centimeters. The width is the length minus 2 centimeters, which is 10 centimeters. Checking the answers for length and width forms the following equation:

$$44 = 2(12) + 2(10)$$

The equation can be solved using the order of operations to form a true statement: $44 = 44$.

29. B: $3x^2 - 3x + 11$. By distributing the implied one in front of the first set of parentheses and the -1 in front of the second set of parentheses, the parenthesis can be eliminated:

$$1(5x^2 - 3x + 4) - 1(2x^2 - 7) = 5x^2 - 3x + 4 - 2x^2 + 7$$

Next, like terms (same variables with same exponents) are combined by adding the coefficients and keeping the variables and their powers the same:

$$5x^2 - 3x + 4 - 2x^2 + 7 = 3x^2 - 3x + 11$$

30. A: Finding the product means distributing one polynomial to the other so that each term in the first is multiplied by each term in the second. Then, like terms can be collected. Multiplying the factors yields the expression:

$$20x^3 + 4x^2 + 24x - 40x^2 - 8x - 48$$

Collecting like terms means adding the x^2 terms and adding the x terms. The final answer after simplifying the expression is:

$$20x^3 - 36x^2 + 16x - 48$$

31. D: The equation can be solved by factoring the numerator into $(x + 6)(x - 5)$. Since that same factor $(x - 5)$ exists on top and bottom, that factor cancels. This leaves the equation $x + 6 = 11$. Solving the equation gives the answer $x = 5$. When this value is plugged into the equation, it yields a zero in the denominator of the fraction. Since this is undefined, there is no solution.

32. B: There are two zeros for the given function. They are $x = 0, -2$. The zeros can be found a number of ways, but this particular equation can be factored into:

$$f(x) = x(x^2 + 4x + 4) = x(x + 2)(x + 2)$$

By setting each factor equal to zero and solving for x, there are two solutions. On a graph, these zeros can be seen where the line crosses the x-axis.

33. C: $y = 40x + 300$. In this scenario, the variables are the number of sales and Karen's weekly pay. The weekly pay depends on the number of sales. Therefore, weekly pay is the dependent variable (y), and the number of sales is the independent variable (x). Each pair of values from the table can be written as an ordered pair (x, y): (2, 380), (7, 580), (4, 460), (8, 620). The ordered pairs can be substituted into the equations to see which creates true statements (both sides equal) for each pair. Even if one ordered pair produces equal values for a given equation, the other three ordered pairs must be checked.

The only equation which is true for all four ordered pairs is $y = 40x + 300$:

$$380 = 40(2) + 300 \rightarrow 380 = 380$$

$$580 = 40(7) + 300 \rightarrow 580 = 580$$

$$460 = 40(4) + 300 \rightarrow 460 = 460$$

$$620 = 40(8) + 300 \rightarrow 620 = 620$$

34. C: The area of the shaded region is the area of the square, minus the area of the circle. The area of the circle will be πr^2. The side of the square will be $2r$, so the area of the square will be $4r^2$. Therefore, the difference is:

$$4r^2 - \pi r^2 = (4 - \pi)r^2$$

35. C: Because the triangles are similar, the lengths of the corresponding sides are proportional. Therefore:

$$\frac{30 + x}{30} = \frac{22}{14} = \frac{y + 15}{y}$$

This results in the equation $14(30 + x) = 22 \times 30$ which, when solved, gives $x = 17.1$. The proportion also results in the equation $14(y + 15) = 22y$ which, when solved, gives $y = 26.3$.

36. B: The technique of completing the square must be used to change $4x^2 + 4y^2 - 16x - 24y + 51 = 0$ into the standard equation of a circle. First, the constant must be moved to the right-hand side of the equals sign, and each term must be divided by the coefficient of the x^2 term (which is 4).

The x and y terms must be grouped together to obtain:

$$x^2 - 4x + y^2 - 6y = -\frac{51}{4}$$

Then, the process of completing the square must be completed for each variable. This gives:

$$(x^2 - 4x + 4) + (y^2 - 6y + 9) = -\frac{51}{4} + 4 + 9$$

The equation can be written as:

$$(x - 2)^2 + (y - 3)^2 = \frac{1}{4}$$

Therefore, the center of the circle is (2, 3) and the radius is:

$$\sqrt{\frac{1}{4}} = \frac{1}{2}$$

37. D: 3 must be multiplied times $27\frac{3}{4}$. In order to easily do this, the mixed number should be converted into an improper fraction.

$$27\frac{3}{4} = \frac{27 * 4 + 3}{4} = \frac{111}{4}$$

Therefore, Denver had approximately:

$$\frac{3x111}{4} = \frac{333}{4} \text{ inches of snow}$$

The improper fraction can be converted back into a mixed number through division.

$$\frac{333}{4} = 83\frac{1}{4} \text{ inches}$$

38.

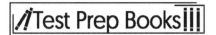

Three girls for every two boys can be expressed as a ratio: 3:2. This can be visualized as splitting the school into 5 groups: 3 girl groups and 2 boy groups. The number of students which are in each group can be found by dividing the total number of students by 5:

$$\frac{650 \text{ students}}{5 \text{ groups}} = \frac{130 \text{ students}}{\text{group}}$$

To find the total number of girls, multiply the number of students per group (130) by the number of girl groups in the school (3). This equals 390.

39. D: $x \leq -5$. When solving a linear equation or inequality:

Distribution is performed if necessary:

$$-3(x + 4) \rightarrow -3x - 12 \geq x + 8$$

This means that any like terms on the same side of the equation/inequality are combined.

The equation/inequality is manipulated to get the variable on one side. In this case, subtracting x from both sides produces $-4x - 12 \geq 8$.

The variable is isolated using inverse operations to undo addition/subtraction. Adding 12 to both sides produces $-4x \geq 20$.

The variable is isolated using inverse operations to undo multiplication/division. Remember if dividing by a negative number, the relationship of the inequality reverses, so the sign is flipped. In this case, dividing by -4 on both sides produces $x \leq -5$.

40.

Kristen bought four DVDs, which would cost a total of $4 \times 15 = \$60$. She spent a total of $100, so she spent $100 - \$60 = \$40$ on CDs. Since they cost $10 each, she must have purchased $40 \div 10 = 4$ CDs.

41.

85% of a number means multiplying that number by 0.85. So, $0.85 \times 20 = \frac{85}{100} \times \frac{20}{1}$, which can be simplified to.

$$\frac{17}{20} \times \frac{20}{1} = 17$$

42. A: To find the fraction of the bill that the first three people pay, the fractions need to be added, which means finding the common denominator. The common denominator will be 60.

$$\frac{1}{5} + \frac{1}{4} + \frac{1}{3} = \frac{12}{60} + \frac{15}{60} + \frac{20}{60} = \frac{47}{60}$$

The remainder of the bill is:

$$1 - \frac{47}{60} = \frac{60}{60} - \frac{47}{60} = \frac{13}{60}$$

43.

The average is calculated by adding all six numbers, then dividing by 6. The first five numbers have a sum of 25. If the total divided by 6 is equal to 6, then the total itself must be 36. The sixth number must be $36 - 25 = 11$.

44. C: Kimberley worked 4.5 hours at the rate of $10/h and 1 hour at the rate of $12/h. The problem states that her pay is rounded to the nearest hour, so the 4.5 hours would round up to 5 hours at the rate of $10/h.

$$(5h) \times \left(\frac{\$10}{h}\right) + (1h) \times \left(\frac{\$12}{h}\right) = \$50 + \$12 = \$62$$

45. C: In order to find the percentage by which the value of the car has been reduced, the current cash value should be subtracted from the initial value and then the difference divided by the initial value. The result should be multiplied by 100 to find the percentage decrease.

$$\frac{20,000 - 8,000}{20,000} = 0.6$$

$$(0.60) \times 100 = 60\%$$

46. A: This problem can be solved by simple multiplication and addition. Since the sale date is over six years apart, 6 can be multiplied by 12 for the number of months in a year, and then the remaining 4 months can be added.

$$(6 \times 12) + 4 = ?$$

$$72 + 4 = 76$$

47. D: This problem can be solved using basic arithmetic. Xavier starts with 20 apples, then gives his sister half, so 20 divided by 2.

$$\frac{20}{2} = 10$$

He then gives his neighbor 6, so 6 is subtracted from 10.

$$10 - 6 = 4$$

Lastly, he uses $\frac{3}{4}$ of his apples to make an apple pie, so to find remaining apples, the first step is to subtract $\frac{3}{4}$ from one and then multiply the difference by 4.

$$\left(1 - \frac{3}{4}\right) \times 4 = ?$$

$$\left(\frac{4}{4} - \frac{3}{4}\right) \times 4 = ?$$

$$\left(\frac{1}{4}\right) \times 4 = 1$$

48. B: To find the value of n, multiply $\frac{5}{2}$ by the reciprocal of $\frac{1}{3}$ and simplify:

$$\frac{5}{2} \div \frac{1}{3} = \frac{5}{2} \times \frac{3}{1} = \frac{15}{2} = 7.5$$

49. A: The total fraction taken up by green and red shirts will be:

$$\frac{1}{3} + \frac{2}{5} = \frac{5}{15} + \frac{6}{15} = \frac{11}{15}$$

The remaining fraction is:

$$1 - \frac{11}{15} = \frac{15}{15} - \frac{11}{15} = \frac{4}{15}$$

50. C: If she has used 1/3 of the paint, she has 2/3 remaining. $2\frac{1}{2}$ gallons are the same as $\frac{5}{2}$ gallons. The calculation is:

$$\frac{2}{3} \times \frac{5}{2} = \frac{5}{3} = 1\frac{2}{3} \; gallons$$

51. C: The volume of a cylinder is $\pi r 2h$, and $\pi \times 6^2 \times 2$ is $72\,\pi$ cm³. Choice *A* is not the correct answer because that is only $6^2 \times \pi$. Choice *B* is not the correct answer because that is $2^2 \times 6 \times \pi$. Choice *D* is not the correct answer because that is $2^3 \times 6 \times \pi$.

52. B: The first step is to calculate the difference between the larger value and the smaller value.

$$378 - 252 = 126$$

To calculate this difference as a percentage of the original value, and thus calculate the percentage *increase*, 126 is divided by 252, then this result is multiplied by 100 to find the percentage: 50%, or Choice *B*.

53. A: Setting up a proportion is the easiest way to represent this situation. The proportion becomes $\frac{20}{x} = \frac{40}{100}$, where cross-multiplication can be used to solve for x. Here, $40x = 2000$, so $x = 50$.

54.

			8	0
⊖	-	-	-	-
	·	·	·	·
⓪	⓪	⓪	●	
①	①	①	①	
②	②	②	②	
③	③	③	③	
④	④	④	④	
⑤	⑤	⑤	⑤	
⑥	⑥	⑥	⑥	
⑦	⑦	⑦	⑦	
⑧	⑧	●	⑧	
⑨	⑨	⑨	⑨	

To solve the problem, a proportion is written consisting of ratios comparing distance and time. One way to set up the proportion is $\frac{3}{48} = \frac{5}{x}$ or $\left(\frac{distance}{time} = \frac{distance}{time}\right)$, where x represents the unknown value of time. To solve a proportion, the ratios are cross-multiplied:

$$(3)(x) = (5)(48)$$

$$3x = 240$$

The equation is solved by isolating the variable, or dividing by 3 on both sides, to produce $x = 80$.

55.

			1	3
⊖	⊖	⊖	⊖	⊖
	○	○	○	○
⓪	⓪		⓪	⓪
①	①		●	①
②	②		②	②
③	③		③	●
④	④		④	④
⑤	⑤		⑤	⑤
⑥	⑥		⑥	⑥
⑦	⑦		⑦	⑦
⑧	⑧		⑧	⑧
⑨	⑨		⑨	⑨

Perimeter is found by calculating the sum of all sides of the polygon. $9 + 9 + 9 + 8 + 8 + s = 56$, where s is the missing side length. Therefore, 43 plus the missing side length is equal to 56. The missing side length is 13 cm.

56.

			3	0
⊖	⊖	⊖	⊖	⊖
	○	○	○	○
⓪	⓪		⓪	●
①	①		①	①
②	②		②	②
③	③		●	③
④	④		④	④
⑤	⑤		⑤	⑤
⑥	⑥		⑥	⑥
⑦	⑦		⑦	⑦
⑧	⑧		⑧	⑧
⑨	⑨		⑨	⑨

When x and y are complementary angles, the sine of x is equal to the cosine of y. The complementary angle of 30 is $90 - 30 = 60$ degrees. Therefore, the answer is 60 degrees.

57.

Start with the original equation: $x - 2xy + 2y$, then replace each instance of x with a 2, and each instance of y with a 3 to get:

$$2^2 - 2 \times 2 \times 3 + 2 \times 3^2$$

$$4 - 12 + 18 = 10$$

58.

Add 3 to both sides to get $4x = 8$. Then, divide both sides by 4 to get $x = 2$.

Dear SAT Test Taker,

We would like to start by thanking you for purchasing this study guide for your SAT exam. We hope that we exceeded your expectations.

Our goal in creating this study guide was to cover all of the topics that you will see on the test. We also strove to make our practice questions as similar as possible to what you will encounter on test day. With that being said, if you found something that you feel was not up to your standards, please send us an email and let us know.

We would also like to let you know about other books in our catalog that may interest you.

ACCUPLACER

This can be found on Amazon: amazon.com/dp/162845945X

AP Biology

amazon.com/dp/1628456221

SAT Math 1

amazon.com/dp/1628458631

TSI

amazon.com/dp/162845699X

We have study guides in a wide variety of fields. If the one you are looking for isn't listed above, then try searching for it on Amazon or send us an email.

Thanks Again and Happy Testing!
Product Development Team
info@studyguideteam.com

FREE Test Taking Tips DVD Offer

To help us better serve you, we have developed a Test Taking Tips DVD that we would like to give you for FREE. **This DVD covers world-class test taking tips that you can use to be even more successful when you are taking your test.**

All that we ask is that you email us your feedback about your study guide. Please let us know what you thought about it – whether that is good, bad or indifferent.

To get your **FREE Test Taking Tips DVD**, email freedvd@studyguideteam.com with "FREE DVD" in the subject line and the following information in the body of the email:

   a. The title of your study guide.

   b. Your product rating on a scale of 1-5, with 5 being the highest rating.

   c. Your feedback about the study guide. What did you think of it?

   d. Your full name and shipping address to send your free DVD.

If you have any questions or concerns, please don't hesitate to contact us at freedvd@studyguideteam.com.

Thanks again!

Made in United States
Orlando, FL
08 April 2022

16607063R00170